Electrical and Mechanical Fault Diagnosis in Wind Energy Conversion Systems

Electrical and Mechanical Fault Diagnosis in Wind Energy Conversion Systems

Edited by

Monia Ben Khader Bouzid
Gérard Champenois

WILEY

First published 2023 in Great Britain and the United States by ISTE Ltd and John Wiley & Sons, Inc.

Apart from any fair dealing for the purposes of research or private study, or criticism or review, as permitted under the Copyright, Designs and Patents Act 1988, this publication may only be reproduced, stored or transmitted, in any form or by any means, with the prior permission in writing of the publishers, or in the case of reprographic reproduction in accordance with the terms and licenses issued by the CLA. Enquiries concerning reproduction outside these terms should be sent to the publishers at the undermentioned address:

ISTE Ltd
27-37 St George's Road
London SW19 4EU
UK

www.iste.co.uk

John Wiley & Sons, Inc.
111 River Street
Hoboken, NJ 07030
USA

www.wiley.com

© ISTE Ltd 2023
The rights of Monia Ben Khader Bouzid and Gérard Champenois to be identified as the authors of this work have been asserted by them in accordance with the Copyright, Designs and Patents Act 1988.

Any opinions, findings, and conclusions or recommendations expressed in this material are those of the author(s), contributor(s) or editor(s) and do not necessarily reflect the views of ISTE Group.

Library of Congress Control Number: 2023938459

British Library Cataloguing-in-Publication Data
A CIP record for this book is available from the British Library
ISBN 978-1-78630-931-0

Contents

Introduction ix
Monia BEN KHADER BOUZID and Gérard CHAMPENOIS

Chapter 1. Accurate Electrical Fault Detection in the Permanent Magnet Synchronous Generator and in the Diode Bridge Rectifier of a Wind Energy Conversion System. 1
Monia BEN KHADER BOUZID and Gérard CHAMPENOIS

    1.1. Introduction. 1
    1.2. Description of the system under study and the used fault detection method 2
    1.3. Fundamental notions of the symmetrical components. 5
    1.4. Development of the analytical expressions of the NSV in the case of the different considered faults 7
        1.4.1. Analytical expression of $\overline{V}_2$ in the case of simultaneous faults . . . 7
        1.4.2. Analytical expression of $\overline{V}_2$ in the case of ITSCF in the PMSG . . 12
        1.4.3. Analytical expression of $\overline{V}_2$ in the case of OCDF in the rectifier . . 14
    1.5. Analytical study of the indicators of the different faults 15
        1.5.1. Analytical study in the case of ITSCF. 16
        1.5.2. Analytical study in the case of OCDF in the rectifier. 19
        1.5.3. Analytical study in the case of SF 24
    1.6. Experimental validation of the proposed fault indicators 25
        1.6.1. Description of the tests process. 25
        1.6.2. Experimental results in the case of healthy operation. 26
        1.6.3. Experimental results in the case of ITSCF in the PMSG. 27
        1.6.4. Experimental results in the case of an OCDF fault in the rectifier . 29
        1.6.5. Experimental results in the case of SF in the system considered . . 31

1.7. Description of the method proposed 32
1.8. Conclusion 37
1.9. References 37

Chapter 2. Control and Diagnosis of Faults in Multiphase Permanent Magnet Synchronous Generators for High-Power Wind Turbines 39
Sérgio CRUZ and Pedro GONÇALVES

2.1. Introduction. 39
2.2. Wind energy conversion systems 40
2.3. Multiphase electric drives on WECS. 41
2.4. Model of a six-phase PMSG drive 43
   2.4.1. Natural reference frame 44
   2.4.2. Synchronous reference frame. 48
2.5. Control strategies 51
   2.5.1. Introduction 51
   2.5.2. Field-oriented control 51
   2.5.3. Direct torque control 52
   2.5.4. Finite control set model predictive control 54
2.6. Fault diagnosis in multiphase drives 71
   2.6.1. Introduction 71
   2.6.2. Interturn short-circuit faults. 73
   2.6.3. High-resistance connections and open-phase faults. 76
   2.6.4. Permanent magnet faults 78
   2.6.5. Current sensor faults 79
   2.6.6. Speed sensor faults 80
2.7. Conclusion 81
2.8. References 82

Chapter 3. Gearbox Fault Monitoring Using Induction Machine Electrical Signals. 89
Khmais BACHA and Walid TOUTI

3.1. Introduction. 89
3.2. Motor stator current signature approach 90
   3.2.1. Air gap magnetic flux density-based approach 90
   3.2.2. Magnetizing current approach 97
3.3. Wound rotor current signature approach. 99
3.4. Experimental results. 101
   3.4.1. MCSA for geared motor fault diagnosis 101
   3.4.2. MCSA for WT gearbox 103

  3.4.3. WT generator current processing. 104
  3.4.4. Current transformations for geared motor fault diagnosis 106
 3.5. Conclusion 116
 3.6. Acknowledgments. 116
 3.7. References 117

Chapter 4. Control of a Wind Distributed Generator for Auxiliary Services Under Grid Faults 119
Youssef KRAIEM and Dhaker ABBES

 4.1. Introduction. 119
 4.2. Description of the renewable distributed generator 123
 4.3. Control of the distributed generator 124
  4.3.1. Control of the wind generator. 124
  4.3.2. Control of the hybrid storage system 128
  4.3.3. Control of the DC bus voltage 130
 4.4. Power management algorithm. 132
  4.4.1. Specifications 132
  4.4.2. Determination of inputs/outputs 133
  4.4.3. Determination of membership functions 133
  4.4.4. Inference engine for energy management. 136
 4.5. Detection and control of the grid faults 138
  4.5.1. Fuzzy logic islanding detection. 141
  4.5.2. Fuzzy droop control technique for the adjustment of the
  grid frequency and voltage 144
 4.6. Simulation results 146
  4.6.1. Control and power management of the distributed generator 147
  4.6.2. Detection and correction of the grid voltage and frequency
  variations at the PCC 150
 4.7. Conclusion 154
 4.8. References 154

Chapter 5. Fault-Tolerant Control of Sensors and Actuators Applied to Wind Energy Systems 159
Elkhatib KAMAL and Abdel AITOUCHE

 5.1. Introduction. 159
 5.2. Objective 161
 5.3. RFFTC of WES with DFIG 163
  5.3.1. TS fuzzy model with parameter uncertainties and
  fuzzy observer 164
  5.3.2. Proposed RFFTC based on FPIEO and FDOS. 167

5.3.3. Proposed RFFTC stability and robustness analysis 170
5.3.4. WES with DFIG application 171
5.3.5. Simulations and results 174
5.4. RFSFTC of WES with DFIG subject to sensor and actuator faults . . . 178
5.4.1. TS fuzzy plant model with actuator faults, sensor faults
and parameter uncertainties. 179
5.4.2. Proposed RFSFTC algorithm based on FPIEO and FDOS 180
5.4.3. Derivation of the stability and robustness conditions 181
5.4.4. WES with DFIG application and simulations and results 183
5.5. RDFFTC of hybrid wind-diesel storage system subject to
actuator and sensor faults 186
5.5.1. Fuzzy observer scheme for the uncertain system with
sensor and actuator faults 187
5.5.2. Proposed RDFFTC, reference model and stability analysis 188
5.5.3. HWDSS application and simulations and results 191
5.6. Conclusion 197
5.7. References 198

List of Authors 203

Index 205

Introduction

Wind energy plays a vital role in meeting the Paris Agreement's goal of 1.5°C global warming and to accelerate the energy transition. In fact, wind energy is a renewable and sustainable source of energy, which does not contribute to greenhouse gas emissions, making it an important tool in combating climate change. As the cost continues to decrease significantly and technology improves, wind energy is becoming more competitive with other sources of energy an increasingly important part of the global energy mix.

According to the Global Wind Energy Council (GWEC), the cumulative capacity of wind power installed worldwide reached 841 GW at the end of 2022. The growth of wind power capacity installation is expected to continue in the coming years as more countries implement policies and invest in renewable energy to reduce their carbon footprint and combat climate change.

A wind energy conversion system is an important technology for generating clean renewable energy and reducing our dependence on fossil fuels. The operating mode of this system consists on capturing the power of the wind and converts it into usable electrical energy. The system typically consists of several key components, including wind turbines composed of blades that capture the kinetic energy of the wind and convert it into rotational motion, the generator that converts the rotational motion of the rotor blades into electrical energy, the power electronic system including inverters, rectifiers and other components that convert the AC power

---

Introduction written by Monia BEN KHADER BOUZID and Gérard CHAMPENOIS.

produced by the generator into a form that can be used by the grid or stored in batteries and the control system responsible for regulating the speed and direction of the rotor blades to optimize the efficiency of the wind turbine.

However, wind energy conversion systems are subject to various types of faults which can impact their reliability and efficiency. These faults can be electrical or mechanical faults. Electrical faults can occur in generators, transformers, power converters and cables. These faults can result in reduced power output, increased maintenance requirements and potentially dangerous situations such as electrical arcing. Mechanical faults can occur in blades, bearings and gears. These faults can result in increased vibration, noise and wear and can ultimately lead to component failure if not addressed.

Therefore, fault detection in wind energy conversion systems is of great important to ensure their reliability, safety and efficiency. Regular maintenance and monitoring can also help to detect them before they lead to downtime or major repairs. Additionally, advanced control and monitoring systems can help to optimize the performance of wind energy conversion systems and reduce the risk of faults occurring.

Thus, this book is an opportunity for readers to deepen their understanding of the theories and concepts related to the topic of electrical and mechanical fault detection and diagnosis in the different components of a wind energy conversion system, as well as to gain insight into the practical applications and the results achieved in the field. To this end, many researchers from the scientific community have contributed to this book in order to share their research results. This book is organized into an Introduction and five chapters.

Chapter 1, *Accurate Electrical Fault Detection in the Permanent Magnet Synchronous Generator and in the Diode Bridge Rectifier of a Wind Energy Conversion System*, written by Monia **Ben Khader Bouzid** and Gérard **Champenois**, proposes an efficient symmetrical component-based method, able to detect, locate and discriminate between an inter-turns short-circuit fault in the permanent magnet synchronous generator and an open-circuit diode fault in the diode rectifier of a small-scale wind conversion energy system. The first part of this chapter will be dedicated to an original analytical study of the negative sequence voltage under the different considered faults, where novel expressions of the negative sequence voltage are developed. Afterward, as a second part, an analytical study of the different proposed indicators of faults will be presented to investigate the behavior of the proposed indicators under the different faulty modes. Then, the third part of this chapter will be focused on the experimental validation of the behavior of the proposed indicators of fault and the novel developed expressions of the negative sequence voltage. Finally, a detailed description of the proposed method will be introduced in the fourth part of this chapter.

Chapter 2, *Control and Diagnosis of Faults in Multiphase Permanent Magnet Synchronous Generators for High-Power Wind Turbines*, written by Sérgio **Cruz** and Pedro **Gonçalves**, presents a general overview of the existing control systems and diagnostic methods available for diagnosing faults in multiphase PMSM drives applied in wind energy conversion systems. After a general overview of the modelling of multiphase PMSM machines, the most common control algorithms of multiphase PMSM drives are presented, including field oriented control, direct torque control and model predictive control (MPC). Special emphasis is given to MPC algorithms due to their increasing popularity and adequacy in the control of this category of drives. Following this, recent diagnostic methods are presented to detect different types of machine and converter faults, including inter-turn short-circuits, high-resistance connections, open-phase faults in the machine and in the power switches, permanent magnet faults, mechanical faults and sensor faults.

Chapter 3, *Gearbox Faults Monitoring Using Induction Machine Electrical Signals*, written by Khmais **Bacha** and Walid **Touti**, first presents the theoretical basis of the AM-FM effect of gear faults on the driven machine stator current using the machine current signal analysis technique (MCSA). Then, the MCSA is compared to various recent methods such as the extended Park vector approach (EPVA) and the discrete cosine/discrete sine transform used for the gear fault diagnosis purpose. Based on the experimental results, these methods are investigated in terms of fault sensitivity to frequency levels.

Chapter 4, *Control of a Wind Distributed Generator for Auxiliary Services Under Grid Faults*, written by Youssef **Kraiem** and Dhaker **Abbes**, presents an intelligent control strategy based on fuzzy logic technology for a renewable distributed generator (RDG) integrated into power electrical system in order to keep the frequency and the voltage of the power grid in an allowable range, while ensuring the continuity of the power supply in the event of a grid fault. The RDG comprises a wind system, as a principal source and a hybrid storage system consisting of battery (BT) and supercapacitors (SC). RDG is associated with loads and a fluctuating power grid. The structure of the proposed control strategy is mainly composed of a fuzzy logic supervisor, a fuzzy detector of the standalone operation mode and an adaptive fuzzy droop control. The fuzzy supervisor is developed to manage the power flows between different sources by choosing the optimal operating mode, while ensuring the stability of the power grid and the continuous supply of loads by maintaining the state of charge of the BT and SC in acceptable levels to improve their lifespans. The fuzzy islanding detector is used to detect the standalone mode in the event of power grid failure. The adaptive fuzzy droop control allows for controlling active and reactive powers exchanged with the power grid, ensuring its stability by maintaining frequency and voltage within optimal margins.

Chapter 5, *Fault-Tolerant Control of Sensors and Actuators Applied to Wind Energy Systems*, written by Elkhatib **Kamal** and Abdel **Aitouche**, proposes an observer-based actuator or sensor detection scheme for TS (Takagi-Sugeno) type fuzzy systems subject to sensor faults, parametric uncertainties and actuator faults. The detection system provides residuals for detecting and isolating sensor faults that may affect a TS model. The fuzzy TS model is adopted for fuzzy modeling of the uncertain nonlinear system and establishing fuzzy state observers. Sufficient conditions are established for robust stabilization in the sense of Lyapunov stability for the fuzzy system. The sufficient conditions are formulated in the form of linear matrix inequality (LMI). The effectiveness of the proposed controller design method is finally demonstrated on a DFIG-based wind turbine to illustrate the effectiveness of the proposed method.

1

Accurate Electrical Fault Detection in the Permanent Magnet Synchronous Generator and in the Diode Bridge Rectifier of a Wind Energy Conversion System

1.1. Introduction

Nowadays, it is a necessity to combat climate change by reducing the emission of greenhouse gases, notably carbon dioxide and methane. Reducing emissions requires essentially generating electricity from renewable energy sources non-emitting carbon. Wind energy is one of the most worldwide used renewable energy, and it is the fastest-growing energy source among the new power generation sources. Wind energy is already rapidly developing into a mainstream power source in many countries of the world, with over 841 GW of installed capacity worldwide (GWEC 2022). This has driven the rapid development and the high-capacity installation of wind energy conversion systems everywhere in the world, onshore and offshore. However, as any system, a wind energy conversion system may be subjected to various types of faults that may negatively affect the continuity of the electrical production and the reliability of such systems. Thus, the implementation of a

---

Chapter written by Monia BEN KHADER BOUZID and Gérard CHAMPENOIS.
For a color version of all figures in this chapter, see www.iste.co.uk/benkhaderbouzid/fault.zip.

monitoring and diagnosis system based on efficient fault detection and diagnosis methods is of great importance to ensure the safety and reliability of a wind energy conversion system.

As the generator and its associated converters are the main electrical components in the energy conversion process of a wind turbine system (Bahloul et al. 2023), this chapter is focused on the detection of the inter-turn short-circuit fault (ITSCF) in the stator windings of the permanent magnet synchronous generator (PMSG) and the open-circuit diode fault (OCDF) in the three-phase diode rectifier connected to the PMSG. Each type of fault can occur separately or two may occur simultaneously. In the case of a simultaneous fault, it is very difficult to discriminate between the two faults. To this end, this chapter proposes an efficient method based essentially on the symmetrical components to detect, locate and discriminate between an ITSCF and an OCDF in a small-scale, variable-speed wind energy conversion system.

This chapter will be organized as follows. The description of the system under study and the principle of the used method is given in section 1.2. The fundamental notions of the symmetrical components are presented in section 1.3 to facilitate the understanding of the presented work. In section 1.4, an original analytical study of the negative sequence voltage (NSV) under the different considered faults is elaborated where new expressions of the NSV under the different considered faults are developed. In section 1.5, the behaviors of the proposed fault indicators are studied analytically. These proposed fault indicators are experimentally validated in section 1.6. The description of the principle of the proposed method using the different fault indicators to detect, discriminate and locate the different considered faults is presented in section 1.7. Finally, a conclusion is drawn in section 1.8.

1.2. Description of the system under study and the used fault detection method

A wind energy conversion system is a cost-effective way to generate clean electricity and protect the environment. This system typically consists of large blades that capture the kinetic energy of the wind and convert it into mechanical power. This mechanical power spins then the generator shaft, which produces electricity. Wind energy conversion systems can be conceived on a small or large scale, onshore or offshore, with fixed or variable speeds and connected or not to the main grid.

In any wind energy conversion system, the key component responsible for the conversion of wind energy to electrical energy is the generator. The typically-used wind turbine generators are a doubly-fed induction generator (DFIG), wound rotor synchronous generator (WRSG), squirrel-cage induction generator (SCIG) and permanent-magnet synchronous generator (PMSG) (Liang et al. 2022). In the last decade, due to its several advantages, the PMSG has gained significant popularity and is becoming widely used in wind energy conversion systems (Yuan et al. 2021). As it has been reported in Gliga et al. (2008), the PMSG faults represent 14.7% of all faults in a WTS, and they account for 24.42% of the downtime.

On the other hand, the wind turbine generator is interfaced with the utility grid via power electronic converters used to transfer and control the wind power into the electric grid. Typically, wind turbine converters include diode rectifier-based converter topology, two-level back-to-back converter topology, three-level neutral-point-clamped back-to-back converter topology and modular multilevel converter topology (Yang et al. 2016). It has been reported in Liang et al. (2022) that the fault of power semiconductor devices is one of the main causes responsible for converter faults. Typical faults of power semiconductor devices can be divided into short-circuit (SC) faults and open-circuit (OC) faults. The OC faults are the most common.

Based on these considerations, this work is interested in the PMSG-based wind turbine structure. However, according to this structure, there are two configurations: PMSG with back-to-back voltage source converters and PMSG with diode rectifiers and boost converters. This chapter is focused on the configuration of the PMSG connected to a diode rectifier-based converter, since this configuration is widely used to interface direct-drive wind conversion energy systems, due to its simplicity and its low cost (Yang et al. 2016). Figure 1.1 shows the block diagram of a small-scale wind conversion system based on a PMSG connected to a diode rectifier. This system is composed of a wind turbine (three blades), a PMSG, a three-phase diode bridge rectifier, a boost chopper and a DC bus. However, the system under study is limited only to the PMSG and the three-phase diode rectifier highlighted in yellow in Figure 1.1.

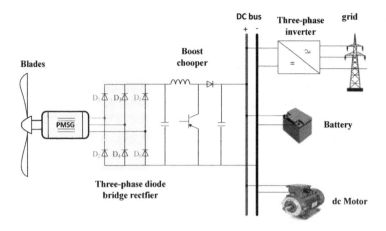

Figure 1.1. *Small-scale, PMSG variable speed wind energy conversion system*

To ensure the reliability and increase the safety of the system under study, this work is aimed at finding an efficient method capable of detecting, locating and discriminating between two electrical faults, which are an inter-turn short-circuit fault (ITSCF) in the stator of the PMSG and an OCDF in the three-phase diode bridge rectifier. This interest is justified by the fact that the ITSCF is the most frequent fault in the machine (Qiao and Lu 2015), and it is a critical and harmful one when detected in the PMSG. It shapes more than one-third of the total faults in the PMSG (Sayed et al. 2021). However, the OCDF in a diode rectifier, although it may not seriously stop the operation of the wind turbine system, can result in overstressing the other healthy diodes and cause the failure of other diodes (Huang et al. 2021).

The proposed fault detection and diagnosis method are based on the monitoring of different relevant indicators of faults extracted basically from the symmetrical components of voltages and currents. It consists first of monitoring the magnitude of the PMSG NSV V_2 to detect any fault in the considered system. Afterward, the mean values of the line currents are used to discriminate between an ITSCF and OCDF since a fault can occur in the PMSG or the rectifier. Furthermore, in the case of an ITSCF or/and an OCDF, the location of the PMSG faulty phase or/and the rectifier's faulty arm is ensured by the monitoring of the phase angle φ_{V2} of the NSV and the phase angle φ_{I2} of the negative sequence current (NSC), respectively. The potential features of these indicators to detect, discriminate and locate the considered faults will be demonstrated in the next sections through an original deep and thorough analytical study of the NSV under the different faults where novel NSV expressions will be developed and presented taking into account the encroachment effect of the current and the placement of the different turns in the slots of the PMSG stator. The behavior of the NSV and the NSC as well as the exactness of the proposed novel expressions will be validated for each fault. However, before investigating the analytical behavior of the NSV under ITSCF and OCDF, let us give first, a brief description of the symmetrical component principle.

1.3. Fundamental notions of the symmetrical components

In a balanced three-phase electrical system composed of a balanced three-phase voltage source connected to a balanced three-phase load, a balanced set of three-phase currents is drawn. In this case, the three-phase voltages are with equal amplitude and 120° spacing. Similarly, the three currents also have equal amplitudes and 120° spacing, but with a phase shift with respect to the voltage due to the lagging or leading loads as illustrated in Figure 1.2.

However, in an unbalanced system, the above principle is no longer true. To simplify the investigation of an unbalanced three-phase system, in 1918 (Furfari et al. 2002), Charles Fortescue suggested that any three-unbalanced phasors can be expressed as the sum of three sets of balanced phasors. Thus, as Fortescue states, an unbalanced three-phase voltage or current system can be decomposed into three symmetrical sets of balanced voltages or currents.

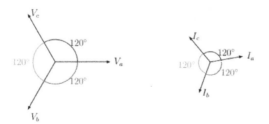

Figure 1.2. *A balanced system of three voltages and currents*

Figure 1.3 illustrates graphically the decomposition of an unbalanced three-phase voltage (Figure 1.3(a)) into three balanced voltages sequences

(Figure 1.3(b)), which $\overline{V}_1$ is called the positive sequence component. It is called positive since it has the same order of phases as in the original set, that is, a → b → c clockwise. $\overline{V}_2$, on the other hand, is referred to as the negative sequence component since it has a flipped order of phases, that is, a → c → b. The last component $\overline{V}_0$ is called the zero sequence component and it consists of three phasors of the same phase shift, that is, a non-rotating set. Similarly, an unbalanced three-phase current system can be decomposed into a positive sequence component $\overline{I}_1$, a negative sequence component $\overline{I}_2$ and a zero sequence component $\overline{I}_0$.

Mathematically, the expressions of the three symmetrical components can be obtained using Fortescue's matrix F:

$$F = \begin{bmatrix} 1 & \overline{a} & \overline{a}^2 \\ 1 & \overline{a}^2 & \overline{a} \\ 1 & 1 & 1 \end{bmatrix} \quad [1.1]$$

Thus, by applying Fortescue's matrix to an unbalance of three-phase voltages $(\overline{V}_1, \overline{V}_2, \overline{V}_3)$ and currents $(\overline{I}_1, \overline{I}_2, \overline{I}_3)$, the expressions of the positive, negative, and zero sequences of voltage and current are given by [1.2] and [1.3], respectively.

$$\begin{bmatrix} \overline{V}_1 \\ \overline{V}_2 \\ \overline{V}_0 \end{bmatrix} = \frac{1}{3} \begin{bmatrix} 1 & \overline{a} & \overline{a}^2 \\ 1 & \overline{a}^2 & \overline{a} \\ 1 & 1 & 1 \end{bmatrix} \cdot \begin{bmatrix} \overline{V}_a \\ \overline{V}_b \\ \overline{V}_c \end{bmatrix} \quad [1.2]$$

$$\begin{bmatrix} \overline{I}_1 \\ \overline{I}_2 \\ \overline{I}_0 \end{bmatrix} = \frac{1}{3} \begin{bmatrix} 1 & \overline{a} & \overline{a}^2 \\ 1 & \overline{a}^2 & \overline{a} \\ 1 & 1 & 1 \end{bmatrix} \cdot \begin{bmatrix} \overline{I}_a \\ \overline{I}_b \\ \overline{I}_c \end{bmatrix} \quad [1.3]$$

where $\overline{a} = e^{j\frac{2\pi}{3}}$ and $\overline{a}^2 = e^{j\frac{4\pi}{3}}$.

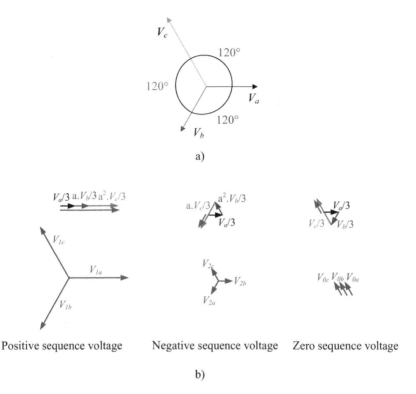

Figure 1.3. *Three-phase unbalanced system: (a) unbalanced voltages; (b) their graphical decomposition into three symmetrical components: positive, negative and zero sequences*

1.4. Development of the analytical expressions of the NSV in the case of the different considered faults

The objective of this novel original analytical study is to develop analytical expressions of the NSV in the three cases of fault, the ITSCF in the machine named machine fault (MF), the OCDF in one arm of the diode full-bridge rectifier named rectifier fault (RF) and the two faults simultaneously named simultaneous faults (SF).

1.4.1. *Analytical expression of $\overline{V}_2$ in the case of simultaneous faults*

To develop the expressions of the NSV in presence of SF, we consider the equivalent electrical circuit illustrated in Figure 1.4. In this circuit, the faulty PMSG presents an ITSCF of N_{sa} shorted turns in phase "a" as an example of ITSCF. This faulty machine is connected to a three-phase diode rectifier with an OCDF modeled by an unbalanced load composed of three unbalanced impedances (Bouzid and Champenois 2017).

The three unbalanced currents flowing in the circuit are denoted $(\overline{I}_a, \overline{I}_b, \overline{I}_c)$. The ITSCF is quantified by the factor "x_a" which is a relative number equal to the ratio between N_{sa} and the total number N_t of turns in the healthy phase, as expressed by [1.4].

$$x_a = N_{sa}/N_t \qquad [1.4]$$

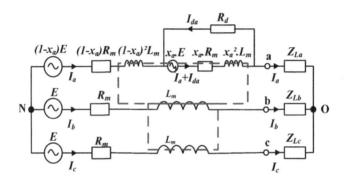

Figure 1.4. *The equivalent electrical circuit of the faulty PMSG with an ITSCF in phase "a"" connected to an unbalanced load*

In general case, for any ITSCF in the phase "i" (i = a, b or c),

$$x_i = N_{si}/N_t \qquad [1.5]$$

According to Figure 1.4, the expressions of the three electromotive forces (EMF) $(\overline{E}_a, \overline{E}_b, \overline{E}_c)$ and the impedance $\overline{Z}_m$ of the healthy PMSG are given by [1.6] and [1.7], respectively.

$$\begin{cases} \overline{E}_a = \sqrt{2}.E.\sin(\omega t) \\ \overline{E}_b = \sqrt{2}.E.\sin(\omega t - 2\pi/3) \\ \overline{E}_c = \sqrt{2}.E.\sin(\omega t + 2\pi/3) \end{cases} \qquad [1.6]$$

$$\bar{Z}_m = R_m + j.w.L_c \quad [1.7]$$

with:

- R_m: phase resistance of the PMSG;
- L_m: self-inductance of the PMSG phase;
- $L_c = L_m \cdot (3/2 - f_t/2)$: cyclic inductance of the PMSG phase; [1.8]
- $\omega = 2.\pi.f$: pulsation of the system;
- f: frequency of the PMSG;
- f_t: leakage flux ratio between phases.

By applying Kirchhoff's voltage law to the equivalent circuit of Figure 1.4, the voltage equations $\bar{V}_{aN}$ $\bar{V}_{bN}$ $\bar{V}_{cN}$, between the neutral point N and the terminals "a", "b" and "c" of the machine, can be written as follows:

$$\bar{V}_{aN} = (1-x_a).\bar{E}_a + x_a.\bar{E}_a - (1-x_a).R_m.\bar{I}_a - j.\omega.[(1-x_a)^2.L_m.\bar{I}_a + M_{ba1}.\bar{I}_b \\ + M_{ca1}.\bar{I}_c + M_{a2a1}.(\bar{I}_{da} + \bar{I}_a)] - x_a.R_m.(\bar{I}_{da} + \bar{I}_a) - j.\omega.\,[x_a^2.L_m.(\bar{I}_{da} + \bar{I}_a) \quad [1.9] \\ + M_{ba2}.\bar{I}_b + M_{ca2}.\bar{I}_c + M_{a1a2}.\bar{I}_a]$$

$$\bar{V}_{bN} = \bar{E}_b - R_m.\bar{I}_b - j.\omega.[L_m \bar{I}_b + M_{a2b}.(\bar{I}_{da} + \bar{I}_a) + M_{alb}.\bar{I}_a + M_{cb}.\bar{I}_c] \quad [1.10]$$

$$\bar{V}_{cN} = \bar{E}_c - R_m.\bar{I}_c - j.\omega.[L_m \bar{I}_c + M_{a2c}.(\bar{I}_{da} + \bar{I}_a) + M_{alc}.\bar{I}_a + M_{bc}.\bar{I}_b] \quad [1.11]$$

With

$$M_{a1a2} = M_{a2a1} = \sqrt{((1-x_a)^2.L_m.x_a^2.L_m)} = x_a.(1-x_a).L_m \quad [1.12]$$

$$M_{a1b} = M_{a1c} = M_{ba1} = M_{ca1} = -(1-f_t)/2.\sqrt{((1-x_a)^2.L_m^2)} \\ = -(1-f_t)/2.(1-x_a).L_m \quad [1.13]$$

$$M_{a2b} = M_{a2c} = M_{ba2} = M_{ca2} = -(1-f_t)/2.\sqrt{(x_a^2.L_m^2)} = -(1-f_t)/2.x_a.L_m \quad [1.14]$$

$$M_{a1b} + M_{a2b} = M_{ba1} + M_{ba2} = M_{a1c} + M_{a2c} = M_{ca1} + M_{ca2}$$
$$= -(1-f_t)/2.L_m.[(1-x_a)+x_a] \quad [1.15]$$
$$= -(1-f_t)/2.L_m$$

$$M_{bc} = M_{cb} = -(1-f_t)/2.L_m \quad [1.16]$$

The expressions of $\overline{V}_{aN}, \overline{V}_{bN}, \overline{V}_{cN}$ can be reformulated according to [1.17] and [1.18].

$$\overline{V}_{aN} = \overline{E}_a - x_a.R_m.\overline{I}_{da} - j.\omega.\overline{I}_{da}.x.L_m - R_m.\overline{I}_a$$
$$-j.\omega.L_m.[\overline{I}_a - (1-f_t)/2.(\overline{I}_b + \overline{I}_c)] \quad [1.17]$$

$$\overline{V}_{bN} = \overline{E}_b + j.\omega.\overline{I}_{da}.x_a.L_m.(1-f_t)/2 - R_m.\overline{I}_b$$
$$-j.\omega.L_m.[\overline{I}_b - (1-f_t)/2.(\overline{I}_c + \overline{I}_a)] \quad [1.18]$$

$$\overline{V}_{cN} = \overline{E}_c + j.\omega.\overline{I}_{da}.x_a.L_m.(1-f_t)/2 - R_m.\overline{I}_c$$
$$-j.\omega.L_m.[\overline{I}_c - (1-f_t)/2.(\overline{I}_a + \overline{I}_b)] \quad [1.19]$$

Since $\overline{I}_b + \overline{I}_c = -\overline{I}_a$, $\overline{I}_c + \overline{I}_a = -\overline{I}_b$ and $\overline{I}_a + \overline{I}_b = -\overline{I}_c$

$$\overline{V}_{aN} = \overline{E}_a - x_a.R_m.\overline{I}_{da} - j.\omega.\overline{I}_{da}.x_a.L_m - R_m.\overline{I}_a$$
$$-j.\omega.L_m.\overline{I}_a.(3/2 - f_t/2) \quad [1.20]$$

$$\overline{V}_{bN} = \overline{E}_b + j.\omega.\overline{I}_{da}.x_a.L_m.(1-f_t)/2 - R_m.\overline{I}_b$$
$$-j.\omega.L_m.\overline{I}_b.(3/2 - f_t/2) \quad [1.21]$$

$$\overline{V}_{cN} = \overline{E}_c + j.\omega.\overline{I}_{da}.x_a.L_m.(1-f_t)/2 - R_m.\overline{I}_c$$
$$-j.\omega.L_m.\overline{I}_c.(3/2 - f_t/2) \quad [1.22]$$

Applying the complex Fortescue's transformer of [1.1] to the voltages $\overline{V}_{aN}, \overline{V}_{bN}, \overline{V}_{cN}$, the expressions of the symmetrical components of these voltages are obtained according to [1.2]. Thus, the expression of the NSV

$\bar{V}_{2a}$ generated by an ITSCF in phase "a" and an OCDF in the diode rectifier simultaneously is given by [1.23]:

$$\bar{V}_{2a} = (\bar{V}_{aN} + a^2.\bar{V}_{bN} + a.\bar{V}_{cN})/3 \quad [1.23]$$

The substituting of [1.20]–[1.22] in [1.23] yields:

$$3.\bar{V}_{2a} = (\bar{E}_a + a^2.\bar{E}_b + a.\bar{E}_c) - x_a.R_m.\bar{I}_{da} - j.\omega.\bar{I}_{da}.x_a.L_m$$
$$+ j.\omega.\bar{I}_{da}.x_a.L_m.(1-f_t)/2.(a^2+a) - R_m.(\bar{I}_a + a^2.\bar{I}_b + a.\bar{I}_c) \quad [1.24]$$
$$- j.\omega.L_m.(3/2 - f_t/2).(\bar{I}_a + a^2.\bar{I}_b + a.\bar{I}_c)$$

with

$$\bar{E}_a + a^2.\bar{E}_b + a.\bar{E}_c = 0 \quad [1.25]$$

$$\bar{I}_a + a^2.\bar{I}_b + a.\bar{I}_c = 3.\bar{I}_{2a} \quad [1.26]$$

$$\bar{I}_{2a} = \bar{I}_{2a\text{-}ITSCF} + \bar{I}_{2\text{-}OCDF} \quad [1.27]$$

$\bar{I}_{2a}$ is the total NSC which is the superposition of the NSC $\bar{I}_{2a\text{-}ITSCF}$ generated by the ITSCF and the NSC $\bar{I}_{2\text{-}OCDF}$ generated by the OCDF.

Using [1.25] and [1.26], [1.24] becomes:

$$3.\bar{V}_{2a} = - x_a.R_m.\bar{I}_{da} + j.\omega.\bar{I}_{da}.x_a.L_m.(1-f_t)/2.(a^2+a+1-1)$$
$$- j.\omega.x_a.L_m.\bar{I}_{da} - 3.R_m.\bar{I}_{2a} - 3.j.\omega.L_m.(3/2-f_t/2).\bar{I}_{2a} \quad [1.28]$$

Since $a + a^2 + 1 = 0$:

$$3.\bar{V}_{2a} = - x_a.R_m.\bar{I}_{da} - j.\omega.x_a.L_m.\bar{I}_{da} - j.\omega.\bar{I}_{da}.x_a.L_m.(1-f_t)/2$$
$$-3.R_m.\bar{I}_{2a} - 3.j.\omega.L_m.(3/2-f_t/2).\bar{I}_{2a} \quad [1.29]$$

$$3.\bar{V}_{2a} = -x_a.(R_m + j.\omega.L_m.(3/2-f_t/2)).\bar{I}_{da} - 3.(R_m + j.\omega.L_m.(3/2-f_t/2)).\bar{I}_{2a} \quad [1.30]$$

$$\bar{V}_{2a} = -(1/3).(x_a.\bar{Z}_m.\bar{I}_{da}) - (\bar{Z}_m.\bar{I}_{2a}) \quad [1.31]$$

1.4.2. Analytical expression of $\bar{V}_2$ in the case of ITSCF in the PMSG

In this case, the considered equivalent electrical circuit is the circuit of Figure 1.4, where the unbalanced load is replaced by three balanced loads ($\bar{Z}_L$) modeling the healthy diode full-bridge rectifier. Thus, with no OCDF, the NSC $\bar{I}_{2\text{-}OCDF}$ is null, and according to [1.27], the total NSC $\bar{I}_{2a} = \bar{I}_{2a\text{-}ITSCF}$. The NSC in the case of ITSCF is caused only by the NSV $\bar{V}_{2a\text{-}ITSCF}$ generated by the unbalance of the three voltages ($\bar{V}_{aN}, \bar{V}_{bN}, \bar{V}_{cN}$) due to the ITSCF. The expression of $\bar{I}_{2a}$ is then:

$$\bar{I}_{2a} = \bar{I}_{2a\text{-}ITSCF} = \bar{V}_{2a\text{-}ITSCF} / \bar{Z}_L \qquad [1.32]$$

Substitution [1.32] in [1.31] yields:

$$\bar{V}_{2a\text{-}ITSCF} = -(1/3).(x_a . \bar{Z}_m . \bar{I}_{da}) - (\bar{V}_{2a\text{-}ITSCF} . \bar{Z}_m / \bar{Z}_L) \qquad [1.33]$$

As Z_L is higher than the machine impedance Z_m, [1.33] can be approximated by:

$$\bar{V}_{2a\text{-}ITSCF} \approx -(1/3).x_a . \bar{Z}_m . \bar{I}_{da} \qquad [1.34]$$

The expressions of the magnitude $V_{2a\text{-}ITSCF}$ and the phase angle $\varphi_{V2a-ITSCF}$ of the NSV $\bar{V}_{2a\text{-}ITSCF}$ are set by [1.35] and [1.36], respectively.

$$V_{2a\text{-}ITSCF} = (1/3) x_a . |Z_m| . |I_{da}| = \frac{1}{3} x_a . I_{da} . (\sqrt{R_m^2 + (L_c . \omega)^2}) \qquad [1.35]$$

$$\varphi_{V2a-ITSCF} = \varphi_{Zm} - 180° + \varphi_{Ida} = \text{artg}(L_c \omega / R_m) - 180° + \varphi_{Ida} \qquad [1.36]$$

We can note here that $V_{2a\text{-}ITSCF}$ and $\varphi_{V2a-ITSCF}$ are insensitive to the load conditions and this is for any ITSCF in any phase "i" ($i = $ a, b or c). However, $V_{2a\text{-}ITSCF}$ depends on the importance of the fault, the magnitude of the faulty current and the frequency, while $\varphi_{V2a-ITSCF}$ depends only on the ratio $L_c.\omega/R_m$ which depends on the frequency and the phase angle of the faulty current.

With the same approach, the analytical expressions $\overline{V}_{2b\text{-}ITSCF}$ in the case of ITSCF in phase "b" and $\overline{V}_{2c\text{-}ITSCF}$ in the case of ITSCF in phase "c" are defined by [1.37] and [1.40], respectively.

$$\overline{V}_{2b\text{-}ITSCF} = -(1/3).a^2.x_b.\overline{I}_{db}.\overline{Z}_m \qquad [1.37]$$

with

$$V_{2b\text{-}ITSCF} = (1/3).x_b.I_{db}.(\sqrt{R_m^2 + (L_c.\omega)^2}) \qquad [1.38]$$

$$\varphi_{V2b\text{-}ITSCF} = \text{artg}(L_c\omega/R_m) - 180° + \varphi_{Idb} \qquad [1.39]$$

$$\overline{V}_{2c\text{-}ITSCF} = -(1/3).a.x_c.\overline{I}_{dc}.\overline{Z}_m \qquad [1.40]$$

with

$$V_{2c\text{-}ITSCF} = (1/3).x_c.I_{dc}.(\sqrt{R_m^2 + (L_c.\omega)^2}) \qquad [1.41]$$

$$\varphi_{V2c\text{-}ITSCF} = \text{artg}(L_c\omega/R_m) - 180° + \varphi_{Idc} \qquad [1.42]$$

In the general case, for an ITSCF in phase "i", the expressions of $V_{2i\text{-}ITSCF}$ and $\varphi_{V2i\text{-}ITSCF}$ are given by [1.43] and [1.44], respectively.

$$V_{2i\text{-}ITSCF} = (1/3)x_i.I_{di}.(\sqrt{R_m^2 + (L_c.\omega)^2}) \; (i = a, b \text{ or } c) \qquad [1.43]$$

$$\varphi_{V2i\text{-}ITSCF} = \varphi_{Zm} - 180° + \varphi_{Idi} \qquad [1.44]$$

The faulty current $\overline{I}_{di}$ (i = a, b, or c) is in phase with the corresponding faulty phase voltage $\overline{V}_i$ ($\overline{V}_i = V_i \angle \varphi_i$) since it flows in a pure resistance. As a result, $\varphi_{Idi} = \varphi_i$ and the expression $\overline{I}_{di}$ will be as follows:

$$\overline{I}_{di} = \sqrt{2}.I_{d\text{-}rms}.e^{j.\varphi_i} \qquad [1.45]$$

Accordingly, [1.44] can be reformulated as follows:

$$\varphi_{V2i-ITSCF} = \varphi_{Zm} - 180° + \varphi_i \quad [1.46]$$

Thus, for an ITSCF in phase "i", the phase angle $\varphi_{V2i-ITSCF}$ of the NSV is as follows:

$$\varphi_{V2a-ITSCF} = \varphi_{Zm} - 180° + 0° = \varphi_{Zm} - 180° \text{ for ITSCF in "a"} \quad [1.47]$$

$$\varphi_{V2b-ITSCF} = \varphi_{Zm} - 180° + 120° = \varphi_{Zm} - 60° \text{ for ITSCF in "b"} \quad [1.48]$$

$$\varphi_{V2c-ITSCF} = \varphi_{Zm} - 180° - 120° = \varphi_{Zm} + 60° \text{ for ITSCF in "c"} \quad [1.49]$$

1.4.3. Analytical expression of $\overline{V}_2$ in the case of OCDF in the rectifier

In this case of fault, the considered equivalent electrical circuit is the one of Figure 1.4, where the faulty machine is replaced by the healthy PMSG. Thus, with the absence of an ITSCF the faulty current $\overline{I}_{da} = 0$ and according to [1.27], $\overline{I}_{2a}$ will be equal to $\overline{I}_{2a-OCDF}$ ($\overline{I}_{2a} = \overline{I}_{2a-OCDF}$). Therefore, according to [1.31], we obtain:

$$\overline{V}_{2a} = \overline{V}_{2a-OCDF} = -\overline{Z}_m \cdot \overline{I}_{2a-OCDF} \quad [1.50]$$

with

$$\overline{I}_{2a-OCDF} = 1/3(\overline{I}_a + a^2 \cdot \overline{I}_b + a \cdot \overline{I}_c) \quad [1.51]$$

The magnitude $V_{2a-OCDF}$ and the phase angle $\varphi_{V2a-OCDF}$ of the NSV are then:

$$V_{2a-OCDF} = |Z_m| \cdot I_{2a-OCDF} = \sqrt{R_m^2 + (L_c \cdot \omega)^2} \cdot I_{2a-OCDF} \quad [1.52]$$

$$\varphi_{V2a-OCDF} = \varphi_{Zm} - 180° + \varphi_{I2a-OCDF} = \text{artg}(L_c \cdot \omega / R_m) - 180° + \varphi_{I2a-OCDF} \quad [1.53]$$

In the general case, for an OCDF in the arm "i", i = a, b or c, the expressions of $V_{2i\text{-}OCDF}$ and $\varphi_{V2i\text{-}OCDF}$ are as follows:

$$V_{2i\text{-}OCDF} = \sqrt{R_m^2 + (L_c.\omega)^2} . I_{2i\text{-}OCDF} \qquad [1.54]$$

$$\varphi_{V2i\text{-}OCDF} = \operatorname{artg}(L_c\omega/R_m) - 180° + \varphi_{I2i\text{-}OCDF} \qquad [1.55]$$

1.5. Analytical study of the indicators of the different faults

After developing the expression of the NSV in the previous section, we study here the behavior of the different indicators of faults that are the amplitude V_2 and the phase angle φ_{V2} of the NSV, the phase angle φ_{I2} of the NSC and the average currents $\langle I_a \rangle$, $\langle I_b \rangle$, $\langle I_c \rangle$ under different operating conditions to point out their potential features useful to detect, discriminate and locate the different faults.

Therefore, to elaborate an analytical study, it is necessary first to calculate the values of these different indicators under different operating conditions. The different variables are calculated using the parameters of the real machine used in the experiment. These parameters are represented in Table 1.1. The leakage flux f_t is considered here as 30% of the total flux of the machine $f_t = 30\%$.

Parameter	Value	Parameter	Value
Rated power (kW)	3	Flux leakage (%)	30
Rated current (A) RMS	13.7	Number of pole pairs	4
Rated torque (Nm)	19.1	Number of slots	36
Rated speed (rpm)	2,250	Turns number/phase	60
Rated voltage (V)	113	Winding number/pole pair	3
Stator resistance (Ω)	0.4	Turns number/slot/phase	5
Self-inductance (mH)	0.63	Number of phases/slot	2

Table 1.1. *Parameters of the used experimental machine*

1.5.1. *Analytical study in the case of ITSCF*

The magnitude $V_{2i\text{-}ITSCF}$ and the phase angle $\varphi_{V2i\text{-}ITSCF}$ of the NSV for an ITSCF in phase "*i*" are calculated analytically under different operating conditions using [1.43] and [1.44]. Note that to obtain correct values of $\varphi_{V2\text{-}ITSCF}$, it is necessary to take into account the physical distribution of the different turns housed in the different slots. As shown in Figure 1.5, which illustrates the mechanical distribution of the windings of a pair pole over a quarter of the used machine, coil a_1 of the winding "a" is ahead of a mechanical angle of 10° and the coil a_3 is behind of –10° in respect to the whole coil "a" under a pole pair.

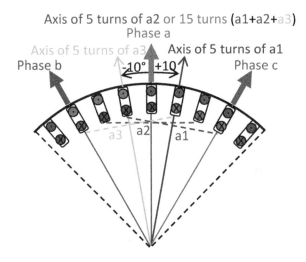

Figure 1.5. *Mechanical distribution of the stator PMSG windings for a pair of poles over a quarter of the machine*

As the machine has four pairs of poles, the electrical phase shift of the EMF of coil a_1 will be 40° ahead of the total EMF of phase "a", which is the sum of coils "a_1", "a_2" and "a_3". The EMF of coil "a_3" will have a phase shift of –40° with respect to the total EMF of phase "a".

Thence, in all this work, it is necessary to add +40° to the voltage angle of 5 turns (a_1), –20° to 10 turns (a_2+a_3) and 0° to 15 turns ($a_1+a_2+a_3$), and the real phase angle will be φ_{V2r} instead φ_{V2}:

$$\varphi_{V2ar\text{-}ITSCF} = \varphi_{V2a\text{-}ITSCF} + 40° \text{ for 5 turns } (a_1)$$

$\varphi_{V2ar\text{-}ITSCF} = \varphi_{V2a\text{-}ITSCF} - 20°$ for 10 turns (a_2+a_3) [1.56]

$\varphi_{V2ar\text{-}ITSCF} = \varphi_{V2a\text{-}ITSCF} + 0°$ for 15 turns $(a_1+a_2+a_3)$

Therefore, the values of V_2, φ_{V2}, φ_{V2r}, and the average currents of the PMSG for different ITSCF in the three phases under different load conditions are represented in Table 1.2. These values are calculated with $I_d = 14$A (RMS) and a frequency $f = 50$ Hz.

	f (Hz)	I_1 (A)	N_s	I_d (A)	φ_{Id} (°)	V_2 (A)	φ_{V2} (°)	φ_{V2r} (°)	$<I_i>$ (A)
Phase a	50	6	5	14	0	0.265	−146	−106	0
		2		14	0	0.265	−146	−106	0
		6	10	14	0	0.53	−146	−166	0
		2		14	0	0.53	−146	−166	0
		6	15	14	0	0.794	−146	−146	0
		2		14	0	0.794	−146	−146	0
Phase b	50	2	5	14	120	0.265	−26	14	0
			10	14	120	0.53	−26	−46	0
			15	14	120	0.794	−26	−26	0
Phase c	50	2	5	14	−120	0.265	94	134	0
			10	14	−120	0.53	94	74	0
			15	14	−120	0.794	94	94	0

Table 1.2. *Analytical calculation of V_2 φ_{V2} and φ_{V2r} for different cases of ITSCF in each PMSG phase under different load conditions*

Based on the values of Table 1.2, it can be noted that:

a) in the case of an ITSCF, the magnitude V_2 of the NSV is non-null, insensitive to the load conditions and is increased with the increasing number of shorted turns. In addition, it can be noted that the values of V_2 are the same for any ITSCF in phases a b or c, which can inform only about the presence of an ITSCF, without informing about the location of the faulty phase. Furthermore, by varying the frequency, as shown in Figure 1.6, V_2 increases with the increase of the frequency. The values of V_2 are the smallest at low frequencies since V_2 is a function of the impedance of the machine Z_m that depends on the frequency as demonstrated by [1.43];

b) φ_{V2r} is insensitive to the load conditions and the severity of the fault, but it depends only on the location of the fault. This is because, for each faulty phase, φ_{V2r} takes a specific value indicating the faulty phase. However, as shown in Figure 1.7, φ_{V2r} varies with the frequency variation since φ_{V2r} depends on φ_{Zm}, which depends in turn on the frequency (see [1.44] and [1.56]). However, as the faulty turns can be situated in coil "a_1" or coil "a_3" as shown in Figure 1.5, φ_{V2r} varies in a range of ±40°. As a result, φ_{V2r} will be situated in a sector of 80° for each faulty phase, which justifies the thick bands of the different curves exhibited in Figure 1.7. Note that the different curves of φ_{V2r} do not present any overlap, which permits us to consider φ_{V2r} as a robust indicator to locate the faulty phase;

c) the three average currents $\langle I_a \rangle, \langle I_b \rangle, \langle I_c \rangle$ are zero.

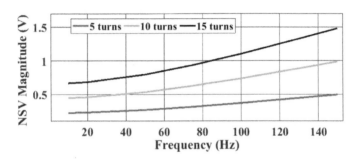

Figure 1.6. *The behavior of $V_{2\text{-}ITSCF}$ under the frequency variation with $I_d = 14A$*

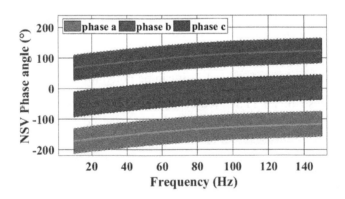

Figure 1.7. *The behavior of $\varphi_{V2ir\text{-}ITSCF}$ in the case of ITSCF in phases "a", "b" and "c" under the frequency variation and the distribution of the faulty turn with $I_d = 14A$*

1.5.2. Analytical study in the case of OCDF in the rectifier

The value of $V_{2i\text{-}OCDF}$ and $\varphi_{V2ir\text{-}OCDF}$ for an OCDF in the arm "i" are calculated using [1.54] and [1.55], respectively. To calculate $V_{2i\text{-}OCDF}$ and $\varphi_{V2ir\text{-}OCDF}$, it is necessary first of all to calculate the three currents $(\bar{I}_a, \bar{I}_b, \bar{I}_c)$ to calculate $\bar{I}_{2i\text{-}OCDF}$ based on [1.3]. However, it is important to mention that the machine's currents are not sinusoidal due to the rectifier, which is a nonlinear load for the PMSG and it is necessary to take into account the encroachment angle μ calculated using [1.56]. This is intended to increase the accuracy of the value of the phase angles of the three fundamental machine currents since these angles are very sensitive to the encroachment angle μ, and the NSV vector (sum of three vectors close to 120°) depends on these angles. Thus, by taking into account μ in the calculation of the fundamental currents, it permits us to increase in consequence the accuracy of the NSV calculation.

$$\mu = (180/\pi).\arccos(1 - 2.(L_c.\omega).I_{dc} / \sqrt{3}.V_{1\max})$$ [1.57]

V_{1max} is the maximum magnitude of the positive sequence voltage (PSV) and I_{dc} is the rectifier output direct current.

To simplify the analytical computation of the PMSG currents, this chapter proposes to linearize the relationship between the different currents and the DC current I_{dc} using an appropriate algorithm illustrated in Figure 1.8. For an open-circuit D_1 taken as an example, this algorithm is based firstly on the application of the FFT on the currents $(\bar{I}_a, \bar{I}_b, \bar{I}_c)$ to obtain their magnitudes and their phase angles $(I_a, \varphi_a, I_b, \varphi_b, I_c, \varphi_c)$ and then on the use of the Fortescue's transform to determine the magnitude and the phase angles of the symmetrical components of the current $(I_{1\text{-}OCDF}, \varphi_{1_1\text{-}OCDF}, I_{2\text{-}OCDF}$ and $\varphi_{1_2\text{-}OCDF})$. It is worth noting here that, for an open-circuit D_1, $i_a(t)$ is entirely negative as shown in Figure 1.8.

Therefore, with I_{dc} = 1A, μ = 10°, and f = 50 Hz, the different ratios provided by the linearization algorithm are denoted:

r_a = 0.550; r_b = 1.156; r_c = 1.13; r_{am} = −0.333; r_{bm} = 0.139; r_{cm} = 0.194, k_1 = 0.915 and k_2 = 0.366, I_2 = 0.4 I_1 [1.58]

The different linear currents can be then expressed as follows:

– For an open-circuit D_1:

$I_a = r_a \cdot I_{dc}$; $I_b = r_b \cdot I_{dc}$; $I_c = r_c \cdot I_{dc}$; $\langle I_a \rangle = r_{am} \cdot I_{dc}$; $\langle I_b \rangle = r_{bm} \cdot I_{dc}$; $\langle I_c \rangle = r_{cm} \cdot I_{dc}$; $I_1 = k_1 \cdot I_{dc}$ and $I_2 = k_2 \cdot I_{dc}$. [1.59]

– For an open-circuit D_3:

$I_a = r_c \cdot I_{dc}$; $I_b = r_a \cdot I_{dc}$; $I_c = r_b \cdot I_{dc}$; $\langle I_a \rangle = r_{cm} \cdot I_{dc}$; $\langle I_b \rangle = r_{am} \cdot I_{dc}$; $\langle I_c \rangle = r_{bm} \cdot I_{dc}$; $I_1 = k_1 \cdot I_{dc}$; and $I_2 = k_2 \cdot I_{dc}$. [1.60]

– For an open-circuit D_5:

$I_a = r_b \cdot I_{dc}$; $I_b = r_c \cdot I_{dc}$; $I_c = r_a \cdot I_{dc}$; $\langle I_a \rangle = r_{bm} \cdot I_{dc}$; $\langle I_b \rangle = r_{cm} \cdot I_{dc}$; $\langle I_c \rangle = r_{am} \cdot I_{dc}$; $I_1 = k_1 \cdot I_{dc}$; and $I_2 = k_2 \cdot I_{dc}$. [1.61]

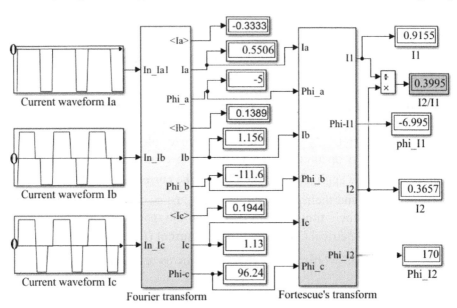

Figure 1.8. Linearization algorithm of the PMSG currents in case of an OCDF of D_1, with $I_{dc} = 1A$ and $\mu = 10°$

To determine the expression of the phase angles φ_a, φ_b and φ_c of the three currents as a function of μ, the algorithm of Figure 1.8 is executed for various values of μ as shown in Table 1.3. From the obtained values, the variation laws of φ_a, φ_b and φ_c as a function of μ have the following expressions given by [1.62]:

$$\varphi_a = \mu/2; \; \varphi_b = (16 - 0.75\mu) - 120; \; \varphi_c = (-16 - 0.75\mu) + 120 \quad [1.62]$$

According to these equations, the calculated phase angles $\varphi_{a\text{-}cal}$, $\varphi_{b\text{-}cal}$ and $\varphi_{c\text{-}cal}$ for different μ are displayed in Table 1.3 to verify the exactness of [1.62]. As shown, their values are exactly equal to the values of φ_a, φ_b and φ_c provided by the algorithm of Figure 1.8, which demonstrates the exactness of the equations of [1.62].

				$-\mu/2$	$(16 - 0.75*\mu) - 120$	$(-16 - 0.75\times\mu) + 120$
μ (°)	φ_a (°)	φ_b (°)	φ_c (°)	φ_a-cal (°)	φ_b-cal (°)	φ_c-cal (°)
2	−1	−105.4	102.4	−1	−105.5	102.5
5	−2.5	−107.8	100.1	−2.5	−107.8	100.3
10	−5	−111.6	96.24	−5	−111.5	96.5
15	−7.5	−115.5	92.44	−7.5	−115.3	92.8
20	−10	−119.4	88.66	−10	−119.0	89.0

Table 1.3. Expressions of the phase angles φ_a, φ_b, φ_c of the three currents as a function of μ with I_{dc} = 1A, f = 100 Hz and an open-circuit fault of D_1

Thereby, using Algorithm 1.1, the magnitude $V_{2i\text{-}OCDF}$ and the phase angle $\varphi_{V2i\text{-}OCDF}$ of the NSV can be calculated. For an open-circuit D_1 (as an example) and for a considered I_{dc}, the EMF of the healthy machine is first calculated (see [1.63]) to calculate $\overline{V}_{1\text{-}OCDF}$ ([1.64]) and its magnitude V_1 ([1.65]). After that, μ is calculated using [1.62] to calculate the currents ($\overline{I}_a, \overline{I}_b, \overline{I}_c, \overline{I}_1, \overline{I}_2$) using [1.66]–[1.70], respectively. Finally, once $\overline{I}_{2\text{-}OCDF}$ [1.70] is calculated, $\overline{V}_{2a\text{-}OCDF}$, $V_{2a\text{-}OCDF}$ and $\varphi_{V2ar\text{-}OCDF}$ can be consequently calculated using [1.71]–[1.73], respectively.

Algorithm 1.1.

$$\overline{E} = 0.1696.\omega : \text{EMF of the healthy machine} \quad [1.63]$$

$$\overline{V}_{1\text{-}OCDF} = \overline{E} - \overline{Z}_m.\overline{I}_{1\text{-}OCDF} \quad [1.64]$$

$$V_{1\text{-}OCDF} = \sqrt{\overline{V}_{1\text{-}OCDF}.(\text{conj}.\overline{V}_{1\text{-}OCDF})} \quad [1.65]$$

$$\overline{I}_{a\text{-}OCDF} = I_a.e^{j.(\mu/2)(\pi/180)} \quad [1.66]$$

$$\overline{I}_{b\text{-}OCDF} = I_b.e^{j((16-0.75\mu).\pi/180 - 2\pi/3)} \quad [1.67]$$

$$\overline{I}_{c\text{-}OCDF} = I_c.e^{j((-16-0.75\mu).\pi/180 + 2\pi/3)} \quad [1.68]$$

$$\overline{I}_{1a\text{-}OCDF} = (1/3).(\overline{I}_{a\text{-}OCDF} + a.\overline{I}_{b\text{-}OCDF} + a^2\overline{I}_{c\text{-}OCDF}) \quad [1.69]$$

$$\overline{I}_{2a\text{-}OCDF} = (1/3).(\overline{I}_{a\text{-}OCDF} + a^2.\overline{I}_{b\text{-}OCDF} + a.\overline{I}_{c\text{-}OCDF}) \quad [1.70]$$

$$\overline{V}_{2a\text{-}OCDF} = -\overline{Z}_m.\overline{I}_{2a\text{-}OCDF} \quad [1.71]$$

$$V_{2a\text{-}OCDF} = \sqrt{\overline{V}_{2a\text{-}OCDF}.(\text{conj}.\overline{V}_{2a\text{-}OCDF})} \quad [1.72]$$

$$\varphi_{V2ar\text{-}OCDF} = (180/\pi).(\text{angle}(\overline{V}_{2a\text{-}OCDF})) \quad [1.73]$$

Therefore, the calculated values of $I_{2i\text{-}OCDF}$, $\varphi_{I2i\text{-}OCDF}$, $V_{2i\text{-}OCDF}$, $\varphi_{V2ir\text{-}OCDF}$, $\langle I_a \rangle$, $\langle I_b \rangle$ and $\langle I_c \rangle$ with $f = 50$ Hz and $I_1 = 2$A and 0.6 A are represented in Table 1.4. The analysis of these values shows that:

a) The values of $V_{2\text{-}OCDF}$ are the same for any OCDF in any arm of the rectifier and increase with the increase of the load current I_1. They also increase with the increase of the frequency as shown in Figure 1.9, since $V_{2\text{-}OCDF}$ is related to Z_m which in turn is related to the frequency.

	I_1 (A)	V_2 (V)	φ_{V2r} (°)	$<I_a>$ (A)	$<I_b>$ (A)	$<I_c>$ (A)	I_2 (A)	φ_{I2} (°)
D_1	2	0.38	24.7	−0.72	0.31	0.42	0.8	171
	0.6	0.115	28.4	−0.22	0.09	0.12	0.24	175
D_2	2	0.38	24.7	0.72	−0.31	−0.42	0.8	171
	0.6	0.115	28.4	0.22	−0.09	−0.12	0.24	175
D_3	2	0.38	125	0.41	−0.72	0.31	0.8	−69
	0.6	0.115	149	0.12	−0.22	0.09	0.24	−65
D_5	2	0.38	−95	0.31	0.41	−0.72	0.8	51
	0.6	0.115	−92	0.09	0.12	−0.22	0.24	55

Table 1.4. Expressions of the phase angles φ_a, φ_b, φ_c of the three currents as a function of µ with I_{dc} = 1A, f = 100 Hz, and an open-circuit fault of D_1

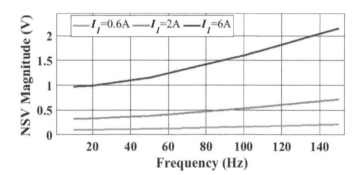

Figure 1.9. The behavior of the magnitude $V_{2\text{-}OCDF}$ of the NSV as a function of the frequency with I_1 = 0.6 A, 2 A and 6 A

b) The values of I_2 are the same for any OCDF and it is always equal to 40% of I_1; $\overline{I_2} = 0.4.I_1.e^{j\varphi_{I2}}$. [1.74]

c) The values of $\varphi_{V2r\text{-}OCDF}$ and $\varphi_{I2\text{-}OCDF}$ depend on the faulty arm without informing about the faulty diode and both of them vary slightly with the load current I_1. However, as shown in Figure 1.10, $\varphi_{V2r\text{-}OCDF}$ varies significantly with the frequency since it is related to φ_{Zm} (see [1.55]), which in turn is related to the frequency, while $\varphi_{I2\text{-}OCDF}$ is almost insensitive to the frequency as seen in Figure 1.11. This aspect lets $\varphi_{I2\text{-}OCDF}$ be more robust to locate the faulty arm even with the frequency variation. Figures 1.10 and 1.11 are plotted with a current range from $I_1 = 0.6$ to 6 A, which justify the width of the curves.

d) The values of the average currents are non-null. This result allows the distinction between an OCDF from an ITSF, which leads to considering a priori the average currents as relevant discriminative indicators.

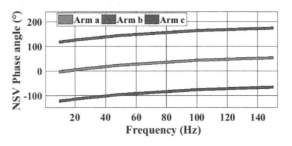

Figure 1.10. *The behavior of the phase angle $\varphi_{V2r\text{-}OCDF}$ of the NSV as a function of the frequency of the NSV with I_1 varying from 0.6 to 6 A*

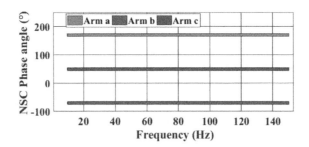

Figure 1.11. *The behavior of the phase angle of the NSC $\varphi_{I2\text{-}OCDF}$ as a function of the frequency with I_1 varying from 0.6 to 6 A*

1.5.3. *Analytical study in the case of SF*

In this case of fault, the magnitude $V_{2\text{-}SF}$ and the phase angle $\varphi_{V2r\text{-}SF}$ of the NSV are calculated using [1.31]. The total NSV $\overline{V}_2$, generated in the considered system, is the vector sum of the NSV $\overline{V}_{2\text{-}ITSCF}$ generated by the MF and the NSV $\overline{V}_{2\text{-}OCDF}$ generated by the RF as given by [1.75].

$$\overline{V}_2 = \overline{V}_{2\text{-}ITSC} + \overline{V}_{2\text{-}OCDF} \qquad [1.75]$$

It has been found that, as in OCDF, $\varphi_{12\text{-}SF}$ is almost insensitive to the load current and the frequency and the values of the average calculated currents have the same values as in the case of OCDF.

1.6. Experimental validation of the proposed fault indicators

1.6.1. *Description of the tests process*

This section aims to validate experimentally the exactness of the developed analytical expressions of the NSV and to verify the behavior of V_2, φ_{V2r}, φ_{I2}, $\langle I_a \rangle$, $\langle I_b \rangle$ and $\langle I_c \rangle$ in cases of ITSCF, OCDF and SF. To this end, different experimental tests are carried out on a test bench illustrated in Figure 1.12.

The used test bench is composed of 36 slots-8 poles, star-connected three-phase surface mounted permanent magnet synchronous machine, driven mechanically by a DC motor. The PMSG parameters are indicated in Table 1.1.

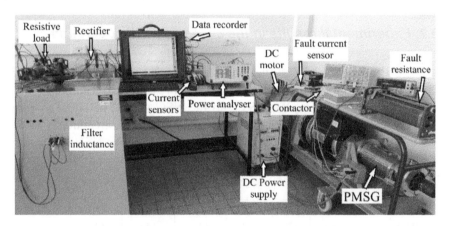

Figure 1.12. *Test bench used in the experimental study*

The stator armature of the PMSG is composed of three concentrated windings with 60 turns per coil distributed among 36 slots. Each winding is made of four coils of 15 turns and each coil is distributed on three slots per pole (two different phases in the same slot). To carry out different stator faults, the stator of the experimental PMSG is equipped with two intermediate taps in the level of the 5th and 15th turns of phase "a". It is possible then to perform ITSCF of 5, 10, 15 and 45 shorted turn in phase "a" using a contactor. For each fault, the tests consist of measuring, with the data recorder, the three voltages ($v_a(t)$, $v_b(t)$, $v_c(t)$) and the four currents ($i_a(t)$, $i_b(t)$, $i_c(t)$ and $i_{da}(t)$) as well as the PMSG frequency with a time acquisition of 2s and a fixed sampling step of 10^{-5} s (100 kHz and 200,000 points).

The processing of the different experimental signals is based firstly on the extraction of the magnitude and the phase angle of each PMSG measured voltages and currents by applying the fast Fourier transformer (FFT) on multiple periods (32 periods to 20 Hz, 64 periods to 50 Hz and 128 periods to 100 Hz) of the measured signal to minimize the errors due to the PMSG inherent asymmetry and second, the calculation of the magnitudes and the phase angles of $\overline{V}_2$ and $\overline{I}_2$ using the Fortescue's transformer (see [1.2]). Note that the calculation of the average experimental currents is performed also with the same periods utilized in the FFT process. The experimental results for the different faults as well as the healthy mode are presented hereafter in the following sections.

1.6.2. *Experimental results in the case of healthy operation*

The experimental values of V_2, φ_{V2r}, $\langle I_a \rangle$, $\langle I_b \rangle$, $\langle I_c \rangle$, I_2 and φ_{I2} in different healthy modes are calculated and represented in Table 1.5. In a perfectly balanced system, the NSV and the NSC should be equal to zero. However, here, it can be noted that in healthy mode the magnitude of the NSV and the NSC is very low. The presence of a low NSV in healthy operation is due to the inherent asymmetry of the machine and the sensor's errors.

f (Hz)	I_1 (A)	V_2 (V)	φ_{V2r} (°)	$\langle I_a \rangle$ (A)	$\langle I_b \rangle$ (A)	$\langle I_c \rangle$ (A)	I_2 (A)	φ_{I2} (°)
20	2.03	0.0002	−119	0.0015	0.0038	−0.0077	0.0002	−53
50	2.02	0.001	−126	−0.001	0.001	−0.003	0.0001	−183
100	2.03	0.0007	−256	0.0002	−0.0043	−0.0057	0.00001	−139

Table 1.5. *Experimental results for different healthy operating modes*

1.6.3. *Experimental results in the case of ITSCF in the PMSG*

The experimental ITSCF tests are carried out in the 5th (a_1), 10th (a_2+a_3) and 15th turns of coil "$a_1+a_2+a_3$" of the winding "a" of the PMSG. The experimental values of V_2, φ_{V2r}, $\langle I_a \rangle$, $\langle I_b \rangle$, $\langle I_c \rangle$, I_2 and φ_{I2} with different cases of ITSCF are calculated and represented in Table 1.6.

f (Hz)	N_{sa}	I_1 (A)	V_2 (V)	φ_{V2r} (°)	$\langle I_a \rangle$ (A)	$\langle I_b \rangle$ (A)	$\langle I_c \rangle$ (A)	I_2 (A)	φ_{I2} (°)
20	5	2.006	0.297	−127	0.0028	0.0016	−0.018	0.0085	−163
	10	1.99	0.447	−189	0.0035	0.0011	−0.011	0.0102	126
50	5	2.004	0.235	−105	−0.0033	0.0001	−0.02	0.0068	−119
	10	2	0.513	−169	−0.0014	0.003	−0.027	0.015	−197
100	5	2.01	0.326	−69	−0.0001	−0.0001	−0.032	0.0045	−107
	10	2.005	0.896	−151	0.001	0.002	−0.03	0.0125	−193

Table 1.6. *Experimental results for different ITSCF in the PMSG with I_{da} = 14A*

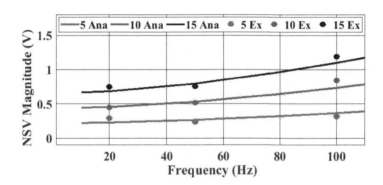

Figure 1.13. *Behavior of the analytical and experimental V_2 under the variation of the frequency for N_{sa} = 5, 10 and 15 shorted turns with I_{da} = 14 A and I_1 = 2 A*

For a fixed frequency, it can be seen that the experimental values of the magnitude V_2 and the phase angle φ_{V2r} of the NSV are extremely close to those of the analytical values given in Table 1.2. In addition, the experimental magnitude V_2 varies slightly with I_1 and the experimental φ_{V2r} indicates correctly the faulty phase, as it has been demonstrated in the analytical study. Furthermore, with the frequency variation, as shown in Figure 1.13, the experimental and the analytical values of NSV magnitude V_2

have the same behavior which proves the dependency of V_2 to the frequency as demonstrated by [1.43].

On the other hand, the experimental values of φ_{V2r} in the case of ITSCF vary with the frequency variation and are situated in the same range of the analytical values as displayed in Figure 1.14.

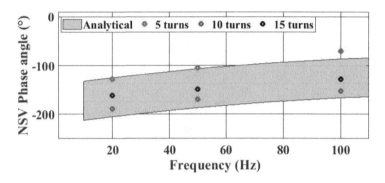

Figure 1.14. *The behavior of the analytical and experimental phase angle φ_{V2r} of the NSV for N_{sa} = 5, 10, and 15 short turns with I_{da} = 14 A and I_1 = 2 A*

The deviation of some points in Figures 1.13 and 1.14 are due to the significant inherent asymmetry and the errors introduced within the calculation of the NSC vector. In fact, according to [1.2], $\overline{V}_2$ is the sum of three vectors $\overline{V}_2 = 1/3(\overline{V}_a + a^2.\overline{V}_b + a.\overline{V}_c)$, important magnitudes (for example 50 V for 50 Hz) and phase angles (close to 120°), the result of this sum is a vector with a small magnitude (around 1 V). Thus, the slightest uncertainty on the amplitudes or the phase shift affects significantly the result $\overline{V}_2$.

Therefore, despite the significant inherent asymmetry of the used PMSG, a good agreement is obtained between the experimental and the analytical results under the variations of the load and the frequency, which leads to the validation of the exactness of the novel proposed analytical expressions of the NSV in case of ITSCF in the PMSG.

Regarding the experimental average currents, Table 1.6 shows that their values are very low and not null as found in the analytical results presented in Table 1.2. This is due to various practical disturbances such as measurement and sensor errors.

1.6.4. Experimental results in the case of an OCDF fault in the rectifier

The experimental PMSG's currents with an open-circuit D_1 and $I_1 = 2A$ are depicted in Figure 1.15 to show their significant distortion due to the OCDF.

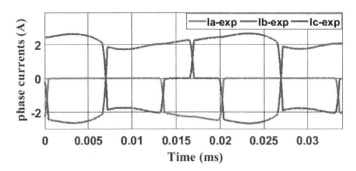

Figure 1.15. Experimental PMSG's currents with an open-circuit D_1 and $I_1 = 2 A$

The experimental values of V_2, φ_{V2r}, I_2, φ_{I2}, $\langle I_a \rangle$, $\langle I_b \rangle$, $\langle I_c \rangle$ and for different OCDF are represented in Table 1.7. It can be noted that, by comparing Table 1.7 to Table 1.4, the experimental and analytical values of the NSV present a good concordance.

f (Hz)	Diode	I_1 (A)	V_2 (V)	φ_{V2r} (°)	$\langle I_a \rangle$ (A)	$\langle I_b \rangle$ (A)	$\langle I_c \rangle$ (A)	I_2 (A)	φ_{I2} (°)
100	D_1	2.07	0.528	40.6	−0.76	0.12	0.65	0.805	167
	D_1	0.65	0.161	29	−0.244	−0.127	0.368	0.246	159
50	D_1	2.07	0.379	20	−0.77	0.127	0.636	0.8	168
	D_1	0.65	0.122	9.1	−0.245	−0.114	0.36	0.243	160
	D_2	0.65	0.124	12	0.243	0.114	−0.358	0.245	160
	D_3	0.65	0.114	133	0.359	−0.24	−0.119	0.246	−79.7
	D_5	0.65	0.119	−99.7	−0.12	0.36	−0.242	0.246	40.2

Table 1.7. Experimental results for different OCDF in the diode rectifier

The magnitude V_2, the phase angles φ_{V2r} and φ_{I2} are also experimentally studied under the variation of the frequency. Thus, as shown in Figures 1.16–1.18, the experimental and analytical V_2, φ_{V2r}, and φ_{I2} present the same behavior under the frequency variation.

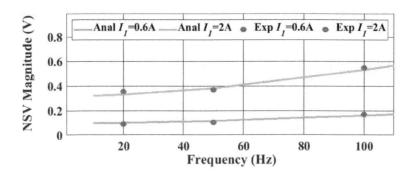

Figure 1.16. The behavior of the experimental and analytical NSV magnitudes in the case of an OCDF (D_1) under the variation of the frequency and the load current

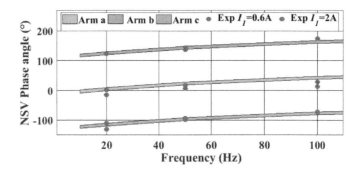

Figure 1.17. The behavior of the experimental and analytical phase angle φ_{V2r} of the NSV in the case of an OCDF (D_1) under the variation of the frequency and the load current

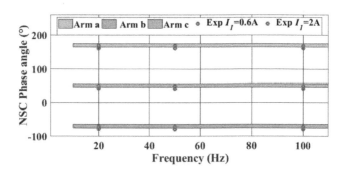

Figure 1.18. The behavior of the experimental and analytical phase angle φ_{I2} of the NSC in the case of an OCDF (D_1) under the variation of the frequency and the load current

As illustrated in Figure 1.16, the behavior of the experimental NSV magnitude follows exactly the evolution of the analytical one, where V_2 increases with the increase of I_1 and the frequency, which confirms that V_2 is related to the frequency as demonstrated in [1.54].

Furthermore, as seen in Figure 1.17, the experimental and the analytical φ_{V2r} are almost situated in the same range. The values of experimental φ_{V2r} also increase with the increase of the frequency and inform about the faulty arm.

Regarding the NSC, the experimental values of the NSC magnitude I_2 are almost equal to the analytical ones given in Table 1.4, with $I_2 = 0.39.I_1$ very close to what was found analytically (see [1.74]). The experimental magnitude I_2 is also insensitive to the frequency variation and they increase only with the increase of I_1. Besides, according to Table 1.7 and Figure 1.18, the experimental φ_{I2} is almost insensitive to the frequency variation as shown in the analytical case in Figure 1.11. Its value informs us of the faulty arm. However, note that the values of the experimental φ_{I2} increases with the increase of I_1 contrary to the analytical φ_{I2}, which decreases with the increase of I_1. This is due to the experimental current ripples shown in Figure 1.15, which were not taken into account in the analytical calculation. This aspect is pointed out especially when the rectified currents ripples are significant in comparison to I_1.

Moreover, by comparing Table 1.7 (case of OCDF) and Table 1.6 (case of ITSCF), the values of the average experimental currents in the case of OCDF are more significant than those with ITSCF. This aspect leads to distinguishing between an ITSCF and an OCDF.

To sum up, the good obtained agreement between the experimental and analytical results validate the exactness of the novel expressions of the NSV, the behavior of φ_{I2} and the three average experimental currents in the case of OCDF.

1.6.5. Experimental results in the case of SF in the system considered

The experimental values of V_2, φ_{V2r}, I_2, φ_{I2} $\langle I_a \rangle, \langle I_b \rangle \langle I_c \rangle$ in the case of SF are represented in Table 1.8 with a faulty current I_{da} equal to 6.5 A for the

ITSCF tests. It can be noted first that the average experimental currents in the case of SF and OCDF are identical. In addition, with the non-informative variation of the values of V_2 and φ_{V2r}, it is impossible to discriminate between an ITSCF and OCDF and in this case, an OCDF can be confused with an SF. Hence, to overcome this issue, this chapter presents a novel technique presented and detailed in the following section.

f (Hz)	N_{sa}	Diode	I_1 (A)	V_2 (V)	φ_{V2r} (°)	$\langle I_a \rangle$ (A)	$\langle I_b \rangle$ (A)	$\langle I_c \rangle$ (A)	I_2 (A)	φ_{I2} (°)
100	5	D_1	2.06	0.475	25.8	−0.761	0.109	0.635	0.802	168
	5	D_1	0.65	0.182	−18.5	−0.244	−0.126	0.361	0.247	160
50	5	D_1	2.07	0.332	3.5	−0.77	0.127	0.631	0.8	168
	5	D_1	0.65	0.13	−44.5	−0.244	−0.118	0.357	0.244	160
	5	D_2	0.65	0.125	−42.5	0.244	0.115	−0.367	0.245	160
	5	D_3	0.65	0.106	−165	0.359	−0.239	−0.12	0.2447	−80
	5	D_5	0.65	0.228	−102	−0.119	0.36	−0.245	0.2431	40

Table 1.8. *Experimental results for different SF with I_{da}= 6.5A for the ITSCF*

1.7. Description of the method proposed

After the above experimental validation, V_2, φ_{V2r}, φ_{I2}, $\langle I_a \rangle, \langle I_b \rangle$ and $\langle I_c \rangle$ are chosen as relevant and robust indicators of the different considered faults. With these reliable indicators, an efficient fault diagnosis method able to detect, locate and discriminate between an ITSCF and OCDF is built. This method is based on three main steps, the detection step, the discrimination step and the location step.

Detection step

Based on Table 1.5, in healthy mode, the values of V_2 are very low under the variation of the frequency and the load current, while in ITSCF (Table 1.6), OCDF (Table 1.7) and SF (Table 1.8), the values of V_2 are more significant. Thus, as illustrated in the flowchart of Figure 1.19, describing the principle of the proposed fault detection, the presence of an NSV V_2 non-null beyond a particular threshold ε_1 ($V_2 > \varepsilon_1$) indicates the existence of a fault in the considered system without specifying its type. Consequently, if

V_2 is less than ε_1 ($V_2 < \varepsilon_1$), the considered system is entirely healthy. Under these considerations, V_2 is chosen as a robust indicator to detect any type of fault in the system under study.

It is worth mentioning that the usage of the threshold ε_1 is to avoid false alarms since even in healthy PMSG operation NSV can be also generated. Thus, from the experimental values of healthy modes under different conditions of frequency and load conditions, ε_1 is fixed to 0.005 V (ε_1 = 0.005 V). This choice is based on the fact that using [1.33] or Figure 1.10, a fault of five shorted turns with I_{da} = 14A and f = 20 Hz generates a V_2 = 0.227 V, thus for a fault of one shorted turn with I_{da} almost equal to 1.5A and f = 20 Hz corresponds to V_2 = 0.005 V, which is the minimum threshold detectable by this algorithm.

Discrimination step between an ITSCF, OCDF and SF

Ideally, in both healthy and ITSCF modes, the values of the average currents are zero because the positive and negative alternations of the current are identical. However, in the real case, in practice, as shown in Tables 1.5 and 1.6, their values are very low different from zero, without exceeding a threshold ε_2. This is due to the measurements and the sensor uncertainties as well as the micro asymmetry of the machine's EMF. On the other hand, the values of the average currents in the case of OCDF are more significant. According to this result, the average currents are chosen as indicators to discriminate effectively between an ITSCF and an OCDF.

Hence, as shown in the flowchart of Figure 1.19, if $\langle I_a \rangle$ or $\langle I_b \rangle$ or $\langle I_c \rangle$ exceeds a threshold ε_2, there is an OCDF; if not, the fault is an ITSCF. From the experimental values, the value of ε_2 is chosen equal to 0.05 A (ε_2 = 0.05 A). This value corresponds to a current I_1 less than 0.2A (which is an overcurrent non-dangerous for the other healthy diode).

It is important to mention that since the values of the average currents in the case of an OCDF are the same as in the case of an SF, based on the average currents, an OCDF may be also an SF which makes it very difficult to discriminate between the two faults easily. Thus, in this work, a novel technique is introduced to distinguish OCDF from an SF. The proposed novel efficient technique consists of the monitoring of $\overline{V}_{2\text{-}ITSC}$ value extracted from [1.75]. The expression of $\overline{V}_{2\text{-}ITSC}$ denoted $\overline{V}_{2rSC}$ is:

$$\overline{V}_{2rSC} = \overline{V}_2 - \overline{V}_{2rD} \qquad [1.75]$$

$$\overline{V}_{2rD} = -\overline{Z}_m \cdot \overline{I}_{2rD} = V_{2rD} \angle \varphi_{V2rD};$$

with $I_{2rD} = 39\%.I_1$, subtracting only the NSV component caused by the OCDF. Thus,

$$\overline{V}_{2rSC} = \overline{V}_2 - \overline{Z}_m.0.39.I_1.e^{j\varphi 12} = V_{2rSC} \angle \varphi_{V2rSC} \qquad [1.76]$$

In this case, as shown in the flowchart of Figure 1.19, if $\overline{V}_{2rSC}$ is less than ε_3, the fault is just an OCDF; otherwise, the fault is an OCDF simultaneously with an ITSCF (SF). This proposed technique, used to discriminate between an OCDF and SF, is also demonstrated experimentally. The experimental results are shown in Table 1.9. Referring to this table, it can be seen that in OCDF, the values of the calculated V_{2rSC} are less than a threshold ε_3 chosen from the experimental values equal to 0.02 V. However, in the case of SF, the values of V_{2rSC} exceed 0.02 V, which means that the fault is an OCDF and an ITSCF simultaneously.

f (Hz)		$Zm.I_2$ (V)		φ_{V2rD} (°)		V_{2rSC} (V)		φ_{V2rSC} (°)	
		OCDF	SF	OCDF	SF	OCDF	SF	OCDF	SF
100	D_1	0.538	0.536	221	221	0.01	0.145	−140	−81.2
	D_1	0.169	0.169	212	213	0.012	0.152	−98.6	−78.4
50	D_1	0.389	0.387	202	202	0.014	0.128	−115	−103
	D_1	0.121	0.121	194	194	0.010	0.123	−74.5	−102
	D_2	0.121	0.121	194	194	0.005	0.117	−38.9	−102
	D_3	0.121	0.121	−46	−46.4	0.008	0.117	−25.9	−99
	D_5	0.122	0.121	73.9	74.1	0.014	0.108	−1.3	−98

Table 1.9. *Experimental results for different SF with I_{da} = 6.5A for the ITSCF*

Note that, in the case of SF, the values of V_{2rSC} are lower than those in the case of ITSCF (Table 1.6), and this is because the tests in SF are performed with a faulty current of 6.5A instead of 14A as in case of ITSCF, in the aim to demonstrate that the proposed method is efficient even with a low faulty current.

Location step

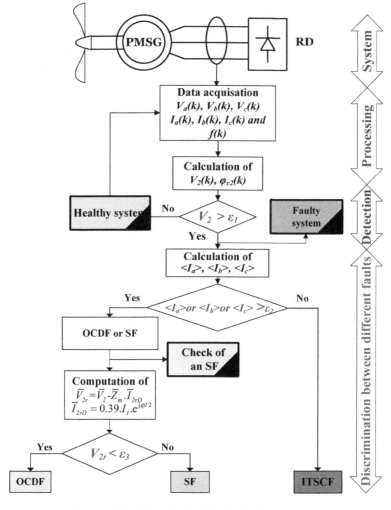

Figure 1.19. *Flowchart of the fault detection and discrimination between ITSCF and OCDF*

For a detected ITSCF or/and OCDF, the location of the faulty phase or/and the faulty arm is based on the monitoring of φ_{V2r} and φ_{I2}.

– For an ITSCF, the phase angle φ_{V2r} of the NSV is chosen as a robust indicator to locate the faulty phase in the case of MF and also in SF. In fact, in SF, it can be noted from Table 1.5 that the phase angle φ_{V2rSC} is situated between −80° for 100 Hz and −100° for 50 Hz, a range corresponding clearly to a fault in phase "a" as shown in Figures 1.7 and 1.15. The flowchart of Figure 1.20 illustrates the principle of the faulty phase location in the case of an MF or SF.

– For an OCDF, the phase angle of NSC φ_{I2} is chosen as a robust indicator to locate the faulty arm (see Table 1.8 and Figure 1.18) instead of φ_{V2r} because it is more robust to the frequency variation. Note also that the sign of the average current of the faulty arm allows for the identification of the open-circuit diode, the top or the bottom one. The flowchart depicted in Figure 1.21 explains clearly the principle of the location of both the faulty arm and the open-circuit diode.

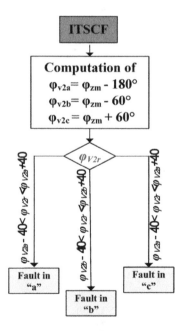

Figure 1.20. *Flowchart of the faulty phase location in case of ITSCF for an MF based on φ_{V2r} or for an SF-based on φ_{V2rSC}*

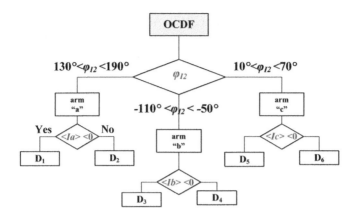

Figure 1.21. *Flowchart of the arm and diode location in case of an OCDF*

1.8. Conclusion

To increase the reliability and safety of a wind energy conversion system, this chapter has focused on the detection, location and discrimination between an ITSCF in a PMSG and an OCDF in a three-phase diode rectifier connected to the PMSG. To this end, a novel and original analytical study of the NSV is thoroughly elaborated, where novel expressions of the NSV under MF, RF and SF are developed and calculated taking into account the position of the shorted turns regarding the slot of the machine and the encroachment effect of the current due to the diode rectifier. Robust and relevant indicators of the considered faults are extracted which are the magnitude V_2 and the phase angle φ_{V2r} of the NSV, the phase angle φ_{12} of the NSC and the three average PMSG currents.

1.9. References

Bahloul, I., Bouzid, B.K.M., Khojet El Khil, S., Champenois, G. (2023). Robust novel indicator to distinguish between an inter-turn short circuit fault and load unbalance in PMSG. *IEEE Transactions on Industry Applications*, 59(3), 3200–3209. doi: 10.1109/TIA.2023.3236898.

Beltran, B., Ahmed-Ali, T., Benbouzid, M.E.H. (2008). Sliding mode power control of variable-speed wind energy conversion systems. *IEEE Transactions on Energy Conversion*, 23(2), 551–558.

Bouzid, B.-K.M. and Champenois, G. (2017). An efficient simplified physical faulty model of permanent magnet synchronous generator dedicated to the stator fault diagnosis – Part II: Automatic stator fault diagnosis. *IEEE Transactions on Industry Applications*, 53(3), 2762–2771.

Furfari, F.A. and Brittain, J. (2002). Charles LeGeyt Fortescue and the method of symmetrical components. *IEEE Industry Applications Magazine*, 8(3), 7–9.

Gliga, L.I., Chafouk, H., Popescu, D., Lupu, C. (2018). Diagnosis of a permanent magnet synchronous generator using the extended Kalman filter and the Fast Fourier Transform. In *7th International Conference on Systems and Control (ICSC)*, 65–70. Valencia.

GWEC (2022). Global wind report [Online]. Available at: https://gwec.net/global-wind-report-2022/.

Huang, W., Du, J., Hua, W., Fan, Q. (2021). An open-circuit fault diagnosis method for PMSM drives using symmetrical and DC components. *Chinese Journal of Electrical Engineering*, 7(3), 124–135.

Liang, J., Zhang, K., Al-Durra, A., Muyeen, S.M., Zhou, D. (2022). A state-of-the-art review on wind power converter fault diagnosis. *Energy Reports*, 8, 5341–5369.

Qiao, W. and Lu, D. (2015). A survey on wind turbine condition monitoring and fault diagnosis Part I: Components and subsystems. *IEEE Transactions on Industrial Electronics*, 62(10), 6536–6545.

Sayed, W.E., Aboelhassan, A., Madi, A., Hebala, A., Galea, M. (2021). Comparative analysis between unscented and extended Kalman filters for PMSG inter-turn fault detection. In *2021 IEEE Workshop on Electrical Machines Design, Control and Diagnosis (WEMDCD)*, 243–248. Modena. doi: 10.1109/WEMDCD51469.2021.9425684.

Yang, Z. and Chai, Y. (2016). A survey of fault diagnosis for onshore grid-connected converter in wind energy conversion systems. *Renewable and Sustainable Energy Reviews*, 66, 345–359.

Yuan, X., Du, Z., Li, Y., Xu, Z. (2021). Control strategies for permanent magnet synchronous generator-based wind turbine with independent grid-forming capability in stand-alone operation mode. *International Transactions on Electrical Energy Systems*, 31(11), 1–22.

2

Control and Diagnosis of Faults in Multiphase Permanent Magnet Synchronous Generators for High-Power Wind Turbines

2.1. Introduction

In recent years, wind turbines (WTs) have been increasingly installed offshore to take advantage of the stronger and steadier wind resources available in these areas (Anaya-Lara et al. 2014). According to the Global Wind Energy Council (GWEC), by the end of 2021, the installed capacity of offshore wind power grew to 57.2 GW, representing a 19.6% growth compared to 2020 (Williams et al. 2022). However, offshore WTs are subjected to more faults and downtime due to accessibility difficulties and harsh environments, which increases their maintenance costs compared to onshore WTs (Anaya-Lara et al. 2018). Hence, developing appropriate control strategies and diagnostic algorithms for offshore WTs is essential to improve their reliability (Faulstich et al. 2011).

This chapter provides a general overview of the existing control systems and diagnostic methods available for diagnosing faults in multiphase permanent magnet synchronous generator (PMSG) drives that can be used in wind energy conversion systems (WECS). After a general description of the modeling of multiphase PMSGs, the most common control algorithms of

---

Chapter written by Sérgio CRUZ and Pedro GONÇALVES.

For a color version of all figures in this chapter, see www.iste.co.uk/benkhaderbouzid/fault.zip.

multiphase PMSG drives are discussed, including field-oriented control, direct torque control and model predictive control (MPC). Particular emphasis is given to MPC algorithms due to their increasing popularity and adequacy in controlling this category of drives. Following this, current diagnostic methods are presented to detect different types of machine and converter faults, including interturn short-circuits, high-resistance connections (HRCs), open-phase faults (OPFs) in the machine and in the power switches, permanent magnet (PM) faults, and sensor faults.

2.2. Wind energy conversion systems

WECS are responsible for converting the kinetic energy of the wind into electrical energy and injecting it into the electric grid. WECS are composed of three types of components (Yaramasu and Wu 2016):

– electrical components: electric generator, power converters, DC-link capacitors, generator-side filters, grid-side filters and step-up transformer;

– mechanical components: tower, nacelle, rotor blades, rotor hub, gearbox, pitch drivers, yaw drives, drivetrain and mechanical brakes;

– control components: controllers for the electrical and mechanical components and associated sensors (e.g. wind speed and wind direction sensors).

At the beginning of the last decade, doubly fed induction generators (DFIG)-based WECS, whose diagram is shown in Figure 2.1, were used in more than 50% of the installed WTs worldwide (Cardenas et al. 2013). However, in recent years there has been a downward trend in the market share of DFIG-based WTs. According to the European Union Science Hub, the DFIG-based WTs represented 34% of the installed European onshore WTs in 2018 (Telsnig 2021). In these systems, variable speed operation of up to ±30% around the synchronous speed is made possible by using partial-rated power converters, with 30% of the DFIG rated power, to connect the rotor windings of the DFIG to the electric grid (Abad et al. 2011). On the other hand, the stator windings of the DFIG are connected to the grid using a step-up transformer. However, DFIG-based WECS require regular maintenance as partial-rated power converters are connected to the rotor windings through slip rings and brushes (Ogidi et al. 2020).

Nowadays, most high-power WTs used in offshore WTs employ WECS based on PMSGs, whose diagram is shown in Figure 2.2 (Yaramasu et al.

2017). In European Union, 80% of offshore WTs installed in 2018 were equipped with PMSG-based WECS (Telsnig 2021). These systems provide a full-speed range of operation (between 0 and 100% of the rated speed), as they use full-rated power converters in a back-to-back configuration with 100% of the PMSG rated power. The main advantages of PMSG-based WTs are increased reliability, higher power density, higher efficiency, lower rotor losses and better voltage/frequency support, making them suitable for application in offshore WTs (Blaabjerg and Ma 2013). Moreover, direct drive PMSG-based WECS do not require a gearbox, which reduces the maintenance requirements of the WT (Zhou et al. 2015). On the other hand, medium-speed PMSG-based WECS require a simpler gearbox and result in a reduction in weight, cost and size of the generator compared to direct drive systems (Moghadam and Nejad 2020).

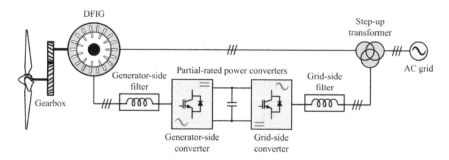

Figure 2.1. *Diagram of a DFIG-based WECS*

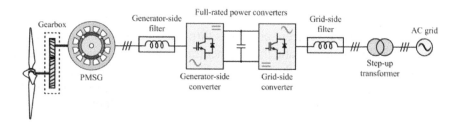

Figure 2.2. *Diagram of a PMSG-based WECS*

2.3. Multiphase electric drives on WECS

Multiphase electric drives is a term widely used in the literature to refer to electric drives based on electric machines with more than three phases, that is, with $m > 3$. Recently, multiphase PMSGs have been applied in offshore WTs to reduce power ratings per phase and improve the fault-tolerant capabilities of WECS (Peng et al. 2021). Independently of the number of phases of an electric machine, the independent control of two currents (two degrees of freedom) is necessary to guarantee a decoupled control of the stator flux and electromagnetic torque. This means it is possible to safely operate an m-phase PMSG with up to m-3 faulty phases without any additional equipment when assuming star-connected windings with a single neutral point (1N). The main advantages of multiphase PMSGs are as follows:

– lower power ratings per phase;

– possibility to operate in fault-tolerant conditions without additional equipment;

– reduced space harmonic content in the magnetomotive force (MMF);

– improved torque profile and efficiency.

In order to maintain compatibility with existing three-phase power converter technology, WECS based on m-phase PMSGs with k sets of three-phase windings, that is, $m = 3.k$, have been proposed for offshore wind energy applications (Figure 2.3) (Prieto-Araujo et al. 2015; Wang et al. 2021). The Envision E128-3.6 MW, the GE Haliade X, and the Vestas V164/174 WTs are some of the several commercial WTs based on six-phase PMSGs ($k = 2$) already available on the market (Cordovil 2018). In these systems, each set of windings is connected to the grid through a group of two-level voltage source inverters (2L-VSIs) in a back-to-back configuration. Each group of 2L-VSIs handles 50% of the PMSG rated power, and typically no current circulates through both groups as each set of windings is wye-connected and the two neutral points are isolated (2N) (Zhu and Hu 2013). Moreover, the two sets of windings are displaced by 30 electrical degrees, usually referred to as an asymmetrical configuration, to reduce the torque harmonic content (Levi et al. 2007; Levi 2008).

Alternatively, the Vensys V70/77 and Goldwind GW70/77 replace the 2L-VSI on the generator side with a diode rectifier and a three-channel boost converter connected to a single DC-link (Yaramasu et al. 2015). However, this configuration forces the circulation of high current harmonics in the generator and does not provide control over the additional degrees of freedom provided by the six-phase PMSG (Xia et al. 2011). Other commercial WTs based on multiphase electric drives are the GE Haliade 150, which is equipped with a nine-phase ($k = 3$) PMSG with 6 MW and back-to-back three-level neutral point clamped converters (3L-NPCs), and the Siemens Gamesa G135 which uses a 12-phase ($k = 4$) PMSG with 5 MW and back-to-back 2L-VSIs (Cordovil 2018).

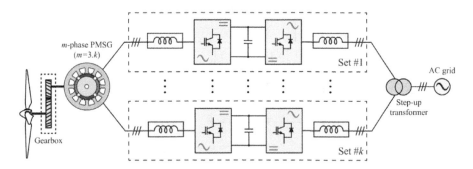

Figure 2.3. *m-Phase PMSG-based WECS with k sets of three-phase windings*

2.4. Model of a six-phase PMSG drive

This section presents the mathematical model of a six-phase PMSG drive, as shown in Figure 2.4. The model presented here can be used to simulate the dynamic behavior of a six-phase PMSG or to develop adequate control strategies and diagnostic techniques.

The model presented in this chapter relies on the following assumptions:

– identical phase windings with a displacement of 30 electrical degrees between the sets of three-phase windings;

– sinusoidally distributed windings, where only the fundamental component of the MMF is considered;

– constant airgap with both saturation and slotting effects neglected;

– iron losses and PM losses are neglected;

– the rotor PMs are mounted on the rotor surface and are radially magnetized;

– motoring convention is selected for current measurement, that is, currents are positive when flowing into the machine terminals.

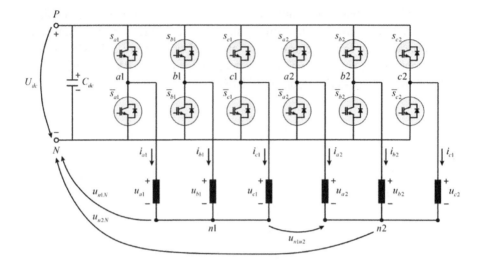

Figure 2.4. *Diagram of the six-phase PMSG drive under study*

2.4.1. Natural reference frame

The dynamic model of an asymmetrical six-phase PMSG (see Figure 2.5 for the sator winding arrangement) with surface-mounted magnets in the natural reference frame (phase variables) is given by:

$$\mathbf{u}_s^{abc} = \mathbf{R}_s \, \mathbf{i}_s^{abc} + \mathbf{L}_s^{abc} \frac{d\mathbf{i}_s^{abc}}{dt} + \mathbf{e}_s^{abc} \qquad [2.1]$$

$$\mathbf{e}_s^{abc} = \frac{d\mathbf{\psi}_{s,PM}^{abc}}{dt} \qquad [2.2]$$

$$\mathbf{\psi}_s^{abc} = \mathbf{L}_s^{abc} \, \mathbf{i}_s^{abc} + \mathbf{\psi}_{s,PM}^{abc}, \qquad [2.3]$$

where $\mathbf{i}_s^{abc}$, $\mathbf{u}_s^{abc}$, $\mathbf{\psi}_s^{abc}$, $\mathbf{\psi}_{s,PM}^{abc}$ and $\mathbf{e}_s^{abc}$ are the vectors of the stator current, voltage, flux linkage, no-load flux linkage and back-EMF. These 6×1 vectors have the following format:

$$\mathbf{f}_s^{abc} = \begin{bmatrix} f_{a1} & f_{b1} & f_{c1} & f_{a2} & f_{b2} & f_{c2} \end{bmatrix}^T \qquad [2.4]$$

Control and Diagnosis of Faults 45

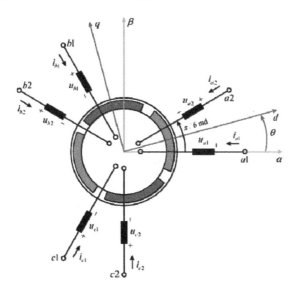

Figure 2.5. *Diagram of asymmetrical PMSM*

The resistance and phase inductance matrices are given by:

$$\mathbf{R}_s = \text{diag}\{R_s, R_s, R_s, R_s, R_s, R_s\} \qquad [2.5]$$

$$\mathbf{L}_s^{abc} = \begin{bmatrix} L_{a1} & M_{a1b1} & M_{a1c1} & M_{a1a2} & M_{a1b2} & M_{a1c2} \\ M_{b1a1} & L_{b1} & M_{b1c1} & M_{b1a2} & M_{b1b2} & M_{b1c2} \\ M_{c1a1} & M_{c1b1} & L_{c1} & M_{c1a2} & M_{c1b2} & M_{c1c2} \\ M_{a2a1} & M_{a2b1} & M_{a2c1} & L_{a2} & M_{a2b2} & M_{a2c2} \\ M_{b2a1} & M_{b2b1} & M_{b2c1} & M_{b2a2} & L_{b2} & M_{b2c2} \\ M_{c2a1} & M_{c2b1} & M_{c2c1} & M_{c2a2} & M_{c2b2} & L_{c2} \end{bmatrix}. \qquad [2.6]$$

By considering the symmetry of the different machine phases, the self and mutual inductances can be related by:

$$L_s = L_{a1} = L_{b1} = L_{c1} = L_{a2} = L_{b2} = L_{c2} \qquad [2.7]$$

$$\begin{aligned} M_s &= M_{a1b1} = M_{a1c1} = M_{b1c1} = M_{c1a1} = M_{c1b1} \\ &= M_{a2b2} = M_{a2c2} = M_{b2c2} = M_{c2a2} = M_{c2b2} \end{aligned} \qquad [2.8]$$

$$\begin{cases} M_1 = M_{a1a2} = M_{b1b2} = M_{c1c2} = M_{a2a1} = M_{b2b1} = M_{c2c1} \\ M_2 = M_{a1b2} = M_{b1c2} = M_{c1a2} = M_{b2a1} = M_{c2b1} = M_{a2c1} = -M_1, \\ M_3 = M_{a1c2} = M_{b1a2} = M_{c1b2} = M_{c2a1} = M_{a2b1} = M_{b2c1} = 0 \end{cases} \quad [2.9]$$

and the inductance matrix $\mathbf{L}_s^{abc}$ can be simplified into:

$$\mathbf{L}_s^{abc} = \begin{bmatrix} L_s & M_s & M_s & M_1 & -M_1 & 0 \\ M_s & L_s & M_s & 0 & M_1 & -M_1 \\ M_s & M_s & L_s & -M_1 & 0 & M_1 \\ M_1 & 0 & -M_1 & L_s & M_s & M_s \\ -M_1 & M_1 & 0 & M_s & L_s & M_s \\ 0 & -M_1 & M_1 & M_s & M_s & L_s \end{bmatrix}, \quad [2.10]$$

where $\{L_s, M_s, M_1\}$ are constant inductances that do not vary with the rotor position.

When considering the no-load flux linkage harmonics due to the rotor PMs, the back-EMF vector is defined as:

$$\mathbf{e}_s^{abc} = -\omega_r \begin{bmatrix} \sum_{h\,\text{odd}}^{\infty} h\psi_{PMh} \sin\left[h(\theta) + \phi_h\right] \\ \sum_{h\,\text{odd}}^{\infty} h\psi_{PMh} \sin\left[h\left(\theta - \frac{2\pi}{3}\right) + \phi_h\right] \\ \sum_{h\,\text{odd}}^{\infty} h\psi_{PMh} \sin\left[h\left(\theta - \frac{4\pi}{3}\right) + \phi_h\right] \\ \sum_{h\,\text{odd}}^{\infty} h\psi_{PMh} \sin\left[h\left(\theta - \frac{\pi}{6}\right) + \phi_h\right] \\ \sum_{h\,\text{odd}}^{\infty} h\psi_{PMh} \sin\left[h\left(\theta - \frac{5\pi}{6}\right) + \phi_h\right] \\ \sum_{h\,\text{odd}}^{\infty} h\psi_{PMh} \sin\left[h\left(\theta - \frac{9\pi}{6}\right) + \phi_h\right] \end{bmatrix}, \quad [2.11]$$

where ψ_{PMh} and ϕ_h are the amplitude and initial phase of the h-order harmonic of the no-load flux linkage, respectively. The rotor electrical speed ω_r and the rotor electrical position θ are related by:

$$\omega_r = \frac{d\theta}{dt}. \qquad [2.12]$$

The electromagnetic torque of a six-phase PMSG is given by Feng et al. (2019):

$$t_e = p \left(\frac{1}{2} \left(\mathbf{i}_s^{abc} \right)^T \frac{d\mathbf{L}_s^{abc}}{d\theta} \mathbf{i}_s^{abc} + \left(\mathbf{i}_s^{abc} \right)^T \frac{d\mathbf{\psi}_{s,PM}^{abc}}{d\theta} \right)\bigg|_{\mathbf{i}_s^{abc}=\text{const.}}, \qquad [2.13]$$

where p is the pole-pair number. Since $\mathbf{L}_s^{abc}$ only contains constant elements, the torque is simply given by Nakao and Akatsu (2014):

$$t_e = p \left(\mathbf{i}_s^{abc} \right)^T \frac{d\mathbf{\psi}_{s,PM}^{abc}}{d\theta} = \frac{p}{\omega_r} \left(\mathbf{i}_s^{abc} \right)^T \mathbf{e}_s^{abc}. \qquad [2.14]$$

Considering ideal semiconductors, the PMSG phase voltages can be related to the switching states of the power switches of the 2L-VSIs using:

$$\mathbf{u}_c^{abc} = U_{dc} \mathbf{M} \, \mathbf{s}, \qquad [2.15]$$

with:

$$\mathbf{M} = \frac{1}{3} \begin{bmatrix} 2 & -1 & -1 & 0 & 0 & 0 \\ -1 & 2 & -1 & 0 & 0 & 0 \\ -1 & -1 & 2 & 0 & 0 & 0 \\ 0 & 0 & 0 & 2 & -1 & -1 \\ 0 & 0 & 0 & -1 & 2 & -1 \\ 0 & 0 & 0 & -1 & -1 & 2 \end{bmatrix} \qquad [2.16]$$

$$\mathbf{s} = \begin{bmatrix} s_{a1} & s_{b1} & s_{c1} & s_{a2} & s_{b2} & s_{c2} \end{bmatrix}^T, \qquad [2.17]$$

where $s_\chi = \{0,1\}$ is the switching state of the phase $\chi = \{a1,...,c2\}$ of the 2L-VSIs.

2.4.2. Synchronous reference frame

The vector space decomposition (VSD) transformation is widely adopted in the literature when modeling six-phase machines and is defined as Zhao and Lipo (1995):

$$\mathbf{T}_{vsd} = \frac{1}{3}\begin{bmatrix} 1 & \cos\left(\frac{-2\pi}{3}\right) & \cos\left(\frac{-4\pi}{3}\right) & \cos\left(\frac{-\pi}{6}\right) & \cos\left(\frac{-5\pi}{6}\right) & \cos\left(\frac{-9\pi}{6}\right) \\ 0 & -\sin\left(\frac{-2\pi}{3}\right) & -\sin\left(\frac{-4\pi}{3}\right) & -\sin\left(\frac{-\pi}{6}\right) & -\sin\left(\frac{-5\pi}{6}\right) & -\sin\left(\frac{-9\pi}{6}\right) \\ 1 & \cos\left(\frac{-4\pi}{3}\right) & \cos\left(\frac{-2\pi}{3}\right) & \cos\left(\frac{-5\pi}{6}\right) & \cos\left(\frac{-\pi}{6}\right) & \cos\left(\frac{-9\pi}{6}\right) \\ 0 & -\sin\left(\frac{-4\pi}{3}\right) & -\sin\left(\frac{-2\pi}{3}\right) & -\sin\left(\frac{-5\pi}{6}\right) & -\sin\left(\frac{-\pi}{6}\right) & -\sin\left(\frac{-9\pi}{6}\right) \end{bmatrix}, \quad [2.18]$$

where the machine variables (currents, flux linkages and voltages) are transformed into the stationary reference frame using:

$$\begin{bmatrix} f_\alpha & f_\beta & f_x & f_y \end{bmatrix}^T = \mathbf{T}_{vsd}\begin{bmatrix} f_{a1} & f_{b1} & f_{c1} & f_{a2} & f_{b2} & f_{c2} \end{bmatrix}^T, \quad [2.19]$$

and subsequently transformed to a synchronous reference frame with:

$$\begin{bmatrix} f_d & f_q & f_{x'} & f_{y'} \end{bmatrix}^T = \mathbf{R}\begin{bmatrix} f_\alpha & f_\beta & f_x & f_y \end{bmatrix}^T, \quad [2.20]$$

with the rotation matrix $\mathbf{R}$ defined as:

$$\mathbf{R} = \begin{bmatrix} \cos(\theta) & \sin(\theta) & 0 & 0 \\ -\sin(\theta) & \cos(\theta) & 0 & 0 \\ 0 & 0 & \cos(\theta) & -\sin(\theta) \\ 0 & 0 & \sin(\theta) & \cos(\theta) \end{bmatrix}. \quad [2.21]$$

The mathematical model of the six-phase PMSG in the synchronous reference frame is given by:

$$\mathbf{u}_s = R_s \mathbf{i}_s + \mathbf{L}_s \frac{d}{dt}\mathbf{i}_s + \omega_r \mathbf{J}\mathbf{L}_s \mathbf{i}_s + \mathbf{e}_s \quad [2.22]$$

$$e_s = \omega_r \mathbf{J} \boldsymbol{\psi}_{s,\text{PM}} + \frac{d\boldsymbol{\psi}_{s,\text{PM}}}{dt} \qquad [2.23]$$

$$\boldsymbol{\psi}_s = \mathbf{L}_s \mathbf{i}_s + \boldsymbol{\psi}_{s,\text{PM}}, \qquad [2.24]$$

where vectors $\mathbf{i}_s$, $\mathbf{u}_s$, $\boldsymbol{\psi}_s$, $\boldsymbol{\psi}_{s,\text{PM}}$ and $\mathbf{e}_s$ have the following format:

$$\mathbf{f} = \begin{bmatrix} f_d & f_q & f_{x'} & f_{y'} \end{bmatrix}^T. \qquad [2.25]$$

The inductance matrix in [2.22] and [2.24] is obtained as:

$$\mathbf{L}_s = 3(\mathbf{R}\,\mathbf{T}_{\text{vsd}})\mathbf{L}_s^{abc}(\mathbf{R}\,\mathbf{T}_{\text{vsd}})^T = \begin{bmatrix} L_{dq} & 0 & 0 & 0 \\ 0 & L_{dq} & 0 & 0 \\ 0 & 0 & L_{xy} & 0 \\ 0 & 0 & 0 & L_{xy} \end{bmatrix} \qquad [2.26]$$

with:

$$L_{dq} = L_s + \sqrt{3}M_1 - M_s, \quad L_{xy} = L_s - \sqrt{3}M_1 - M_s. \qquad [2.27]$$

Matrix $\mathbf{J}$ in [2.22]–[2.23] is given by:

$$\mathbf{J} = \frac{1}{\omega_r}(\mathbf{R})\frac{d(\mathbf{R})^{-1}}{dt} = \begin{bmatrix} 0 & -1 & 0 & 0 \\ 1 & 0 & 0 & 0 \\ 0 & 0 & 0 & 1 \\ 0 & 0 & -1 & 0 \end{bmatrix}. \qquad [2.28]$$

Considering the harmonics until the 13th, the back-EMF vector $\mathbf{e}_s$ is given by:

$$\mathbf{e}_s = \omega_r \begin{bmatrix} -11\psi_{\text{PM}11}\sin(12\theta + \phi_{11}) - 13\psi_{\text{PM}13}\sin(12\theta + \phi_{13}) \\ \psi_{\text{PM}1} - 11\psi_{\text{PM}11}\cos(12\theta + \phi_{11}) + 13\psi_{\text{PM}13}\cos(12\theta + \phi_{13}) \\ -5\psi_{\text{PM}5}\sin(6\theta + \phi_5) - 7\psi_{\text{PM}7}\sin(6\theta + \phi_7) \\ 5\psi_{\text{PM}5}\cos(6\theta + \phi_5) - 7\psi_{\text{PM}7}\cos(6\theta + \phi_7) \end{bmatrix}. \qquad [2.29]$$

The electromagnetic torque produced by the six-phase PMSG is given by:

$$t_e = \frac{p}{\omega_r} \left(3(\mathbf{R}\,\mathbf{T}_{vsd})^T \mathbf{i}_s\right)^T \left(3(\mathbf{R}\,\mathbf{T}_{vsd})^T \mathbf{e}_{s,PM}\right)$$
$$= 3\frac{p}{\omega_r}\left(e_d i_d + e_q i_q + e_{x'} i_{x'} + e_{y'} i_{y'}\right)$$

[2.30]

Since the amplitude of ψ_{PM1} is usually much larger than the other harmonic components of the no-load flux in [2.29], the torque developed by the PMSG can be controlled by regulating the q-axis current according to [2.30].

The PMSG voltages in the stationary reference frame are given by:

$$\mathbf{u}_s^{\alpha\beta} = \mathbf{T}_{vsd} \cdot \mathbf{u}_s^{abc}$$
$$= \begin{bmatrix} u_\alpha & u_\beta & u_x & u_y \end{bmatrix}$$

[2.31]

where by considering the $2^6 = 64$ switching states of the 2L-VSIs, the mapping of the different vectors in the α-β and x-y subspaces is presented in Figure 2.6.

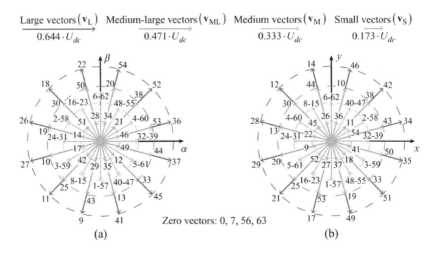

Figure 2.6. *Voltage vectors mapping into the (a) α – β subspace and (b) x-y subspace*

2.5. Control strategies

2.5.1. *Introduction*

The control of PMSG-based WECS is typically divided into two control subsystems: (a) control of the generator-side converters and (b) control of the grid-side converters. The first one is responsible for regulating the generator currents, power, speed or torque and flux, whereas the second one is responsible for controlling the DC-link voltage and the active and reactive powers injected into the grid. The control of the generator-side converters is responsible for guaranteeing a high-dynamic performance and efficient operation of the PMSG, and it is the main focus of this section. The most commonly used control strategies for generator-side converters are field-oriented control (FOC) and direct torque control (DTC). Furthermore, finite control set model predictive control (FCS-MPC) has emerged as an interesting alternative to classical control strategies and is also included in this section. More information on the control of grid-side inverters can be found in Anaya-Lara et al. (2014, 2018).

2.5.2. *Field-oriented control*

The main objective of FOC is to provide decoupled control of the machine stator flux and electromagnetic torque. In the case of six-phase PMSGs, this strategy controls the stator flux of the machine in a synchronous reference frame, which is aligned with the no-load flux linkage due to the rotor PMs (Duran et al. 2017). The diagram of a typical FOC strategy for six-phase PMSG drives is shown in Figure 2.7. The *d-q* current components are regulated by PI controllers to adjust the stator flux and torque, respectively, while the *x'-y'* currents are adjusted by PI and PR controllers associated in parallel (Eldeeb et al. 2019). Although the *x'-y'* currents do not contribute to the electromechanical energy conversion, they contribute to the machine losses and should therefore be maintained at zero during healthy and balanced operating conditions. These undesirable current components appear due to the nonlinearities of the power converter switches (e.g. dead-time effects), back-EMF harmonics and small asymmetries (Karttunen et al. 2017). The PR controllers shown in Figure 2.7 are typically tuned at $6\omega_r$ to eliminate the fifth- and seventh-order harmonics from phase currents. These harmonics are mapped as a sixth harmonic in the synchronously rotating (clockwise) *x'-y'* reference frame.

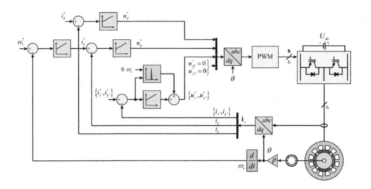

Figure 2.7. *Diagram of a FOC strategy for six-phase PMSG drives*

2.5.3. Direct torque control

The main objective of DTC is to directly control the stator flux and the electromagnetic torque. Unlike FOC, DTC typically controls the stator flux in the stationary reference frame and does not require any rotational transformation. Both the stator flux and electromagnetic torque are regulated with hysteresis controllers, which provides better transient response compared to FOC strategies. A diagram of a DTC strategy for six-phase PMSGs is presented in Figure 2.8, where virtual vectors are considered in the switching table (Ren and Zhu 2015).

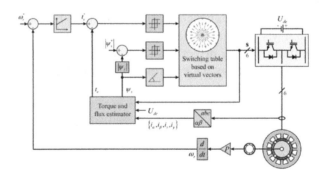

Figure 2.8. *Diagram of a DTC strategy for six-phase PMSG drives*

Since the voltage vectors in six-phase machines (Figure 2.6) are simultaneously mapped into both subspaces, applying a single voltage vector during a control sampling period can result in undesirable *x-y* voltages being applied to the machine. This, in turn, can lead to significant *x-y* currents due to the low equivalent impedance of the machine in this subspace. To overcome this issue, the concept of virtual vectors was first introduced in Zheng et al. (2011). Virtual vectors are created by applying a large vector and a medium-large vector with coincident phases in the $\alpha-\beta$ subspace during T_s with the following duty cycles:

$$d_L = (\sqrt{3}-1), \quad d_{ML} = 1 - d_L, \quad \quad [2.32]$$

where $\{d_L, d_{ML}\}$ are the duty cycles of the large and medium-large vectors, respectively. As an example, virtual vector $\mathbf{v}_{v1}$ is the result of applying the large vector $\mathbf{v}_{36}$ during $d_L T_s$ and the medium-large vector $\mathbf{v}_{53}$ during $d_{ML} T_s$:

$$\begin{aligned}\mathbf{v}_{v1} &= \mathbf{v}_{36} d_L + \mathbf{v}_{53} d_{ML} \\ &= U_{dc}\left[\tfrac{\sqrt{3}}{3} \quad \tfrac{2\sqrt{3}}{3}-1 \quad 0 \quad 0 \quad 0 \quad 0\right]^T.\end{aligned} \quad [2.33]$$

From [2.33], vector $\mathbf{v}_{v1}$ has an amplitude of $2\sqrt{1/3(2-\sqrt{3})} \cdot U_{dc} \approx 0.597 \cdot U_{dc}$ in the $\alpha-\beta$ subspace and zero components in the *x-y* subspace. Figure 2.9 shows the twelve active virtual vectors $\{\mathbf{v}_{v1},...,\mathbf{v}_{v12}\}$. Although the amplitude of virtual vectors in the $\alpha-\beta$ subspace is 7.3% lower than the one of the large vectors (Figure 2.9), they allow a significant reduction of the *x-y* current harmonics.

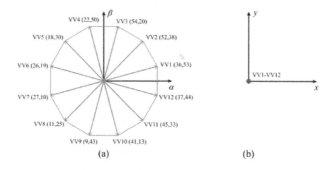

Figure 2.9. *Mapping of the virtual vectors into the (a) $\alpha-\beta$ subspace and (b) x-y subspace*

2.5.4. *Finite control set model predictive control*

Since the last decade, FCS-MPC has been widely researched as an alternative to FOC and DTC to control electric drives. FCS-MPC provides better dynamic performance, simple design, flexibility in the definition of control objectives, easy inclusion of constraints, multivariable control and fewer tuning requirements (Yaramasu and Wu 2016). However, the application of FCS-MPC to commercial drives still faces some challenges, such as higher computational burden and dependence on model parameter accuracy to achieve good control performance (Geyer 2016). The main steps of a typical FCS-MPC strategy can be summarized as:

1) sample the measured signals (e.g. DC-link voltage, phase currents, and rotor electrical position);

2) apply the optimal control actuation selected in the previous sampling period;

3) predict the controlled variables, using a discretised system model, up to two samples ahead;

4) minimize the cost function containing the control objectives to obtain the optimal control actuation.

Different control objectives can be defined in the cost function of FCS-MPC strategies when considering the control of electric drives, such as reference tracking of current, torque flux or speed (Yaramasu and Wu 2016).

Predictive current control (PCC) is one of the most reported variants of FCS-MPC in the literature, and its main objective is to guarantee the reference tracking of the d-, q-, x'- and y'-axes currents (Gonçalves et al. 2019a). The diagram of a standard PCC (S-PCC) strategy for six-phase PMSG drives is shown in Figure 2.10.

However, S-PCC strategies only consider the application of a single voltage vector, with components on both d-q and x'-y' subspaces, which does not lead to an optimal tracking of the d-q and x'-y' currents. Furthermore, since the equivalent impedance of the machine in the x-y subspace is typically low in six-phase machines, the application of any voltage vector with non-zero x-y components leads to increased current harmonics (Gonçalves et al. 2019a). To solve this problem, PCC based on virtual vectors (VV-PCC) was first proposed by Gonzalez-Prieto et al.

(2018) to guarantee the application of average zero x-y voltages to the six-phase machine over a control period. The main objective of VV-PCC is to select the optimal virtual vector that minimizes the reference tracking errors of the d-q currents. However, since virtual vectors have zero components in the x-y subspace, the VV-PCC strategy leaves the x'-y' currents of the machine in open-loop (Duran et al. 2017). In practice, this means that considerable x'-y' current harmonics can circulate in the PMSG windings due to dead-time effects, back-EMF harmonics and machine asymmetries, which increase copper losses in the machine (Karttunen et al. 2017; Gonçalves et al. 2019a).

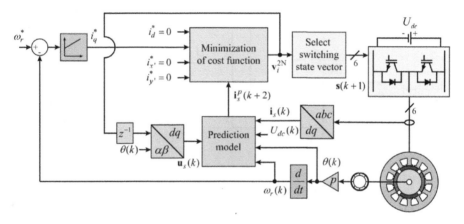

Figure 2.10. *Diagram of the S-PCC strategy for six-phase PMSG drives*

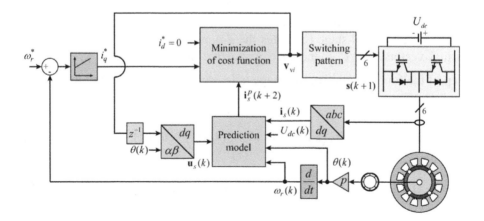

Figure 2.11. *Diagram of the VV-PCC strategy for six-phase PMSG drives*

2.5.4.1. *Finite control set model predictive control*

To overcome the limitations of the VV-PCC strategy in terms of minimizing the *x'-y'* currents in six-phase PMSG drives, a multistage VV-PCC strategy, as shown in Figure 2.12, was proposed by Gonçalves et al. (2022b).

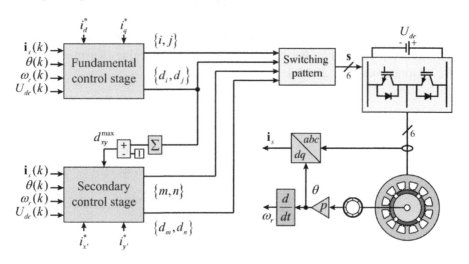

Figure 2.12. *Diagram of the MSVV-PCC strategy for six-phase PMSG drives*

The MSVV-PCC strategy from Figure 2.12 is split into two distinct control stages, a fundamental control stage to regulate the *d-q* currents and a secondary control stage to handle the control of *x'-y'* currents. The fundamental control stage regulates the *d-q* currents and selects two optimal virtual vectors $\{\mathbf{v}_{vi}, \mathbf{v}_{vj}\}$ and their respective duty cycles $\{d_i, d_j\}$. The first step of the fundamental control stage is to predict the *d-q* currents for instant $k+1$ using:

$$i_d^p(k+1) = \left(1 - \frac{R_s T_s}{L_{dq}}\right) i_d(k) + \omega_r(k) T_s i_q(k) + \frac{T_s}{L_{dq}}\left(u_d(k) - e_d(k)\right) \quad [2.34]$$

$$i_q^p(k+1) = \left(1 - \frac{R_s T_s}{L_{dq}}\right) i_q(k) - \omega_r(k) T_s i_d(k) + \frac{T_s}{L_{dq}}\left(u_q(k) - e_q(k)\right), \quad [2.35]$$

where from [2.29], the d-q back-EMF components, up to the seventh-order harmonic, are given by:

$$e_d(k) = 0, \quad e_q(k) = \omega_r(k)\psi_{\text{PM1}}. \qquad [2.36]$$

The predictions of the d-q currents for instant $k+2$, due to the application of an c-index virtual vector $\mathbf{v}_{vc}$, are calculated by:

$$i_{d,vc}^p(k+2) = \left(1 - \tfrac{R_s T_s}{L_{dq}}\right) i_d^p(k+1) + \omega_r(k)\, T_s\, i_q^p(k+1) + \tfrac{T_s}{L_{dq}}\left(u_d(k+1) - e_d(k+1)\right) \qquad [2.37]$$

$$i_{q,vc}^p(k+2) = \left(1 - \tfrac{R_s T_s}{L_{dq}}\right) i_q^p(k+1) - \omega_r(k)\, T_s\, i_d^p(k+1) + \tfrac{T_s}{L_{dq}}\left(u_q(k+1) - e_q(k+1)\right) \qquad [2.38]$$

where the back-EMF components are given by $e_d(k+1) = e_d(k)$, and $e_q(k+1) = e_q(k)$ by assuming the rotor speed remains invariant over a sampling period. The d- and q- voltages for instant $k+1$ are obtained by:

$$\begin{bmatrix} u_d(k+1) & u_q(k+1) & 0 & 0 \end{bmatrix}^T = \mathbf{R} \cdot \mathbf{v}_{vc}, \quad a \in \{1,\ldots,12\}. \qquad [2.39]$$

The fundamental stage determines the indexes $\{i,j\}$ of the optimal pairs of virtual vectors $\{\mathbf{v}_{vi}, \mathbf{v}_{vj}\}$ with:

$$\{i,j\} = \arg\min_{\{a,b\}}(g_{f,a} + g_{f,b}), \quad \{a,b\} = \{\{1,2\},\{2,3\},\ldots,\{12,1\}\}, \qquad [2.40]$$

where $\{g_{f,a}, g_{f,b}\}$ are the cost functions due to the application of $\{\mathbf{v}_{va}, \mathbf{v}_{vb}\}$, which are neighbors in the α-β subspace (Figure 2.6), and are given by:

$$g_{f,a} = \left(i_d^* - i_{d,va}^p(k+2)\right)^2 + \left(i_q^* - i_{q,va}^p(k+2)\right)^2 \qquad [2.41]$$

$$g_{f,b} = \left(i_d^* - i_{d,vb}^p(k+2)\right)^2 + \left(i_q^* - i_{q,vb}^p(k+2)\right)^2 \qquad [2.42]$$

where the d-q current predictions are calculated with [2.37]–[2.38] for the candidate virtual vectors $\{\mathbf{v}_{va}, \mathbf{v}_{vb}\}$.

After determining the indexes $\{i, j\}$ of the optimal pair of virtual vectors with [2.40], the fundamental stage calculates the optimal duty cycles of these vectors $\{d_i, d_j\}$ to minimize the tracking error of the d-q currents. A modulated virtual vector is defined as:

$$\mathbf{v}_{vc,1} = d_{i1} \cdot \mathbf{v}_{vi} + d_{j1} \cdot \mathbf{v}_{vj}, \qquad [2.43]$$

with $\mathbf{v}_{vc,1}$ located inside the dodecagon in Figure 2.9, the duty cycles $\{d_{i1}, d_{j1}\}$ being calculated by minimizing [2.44], subjected to [2.45]–[2.46]:

$$g_{f,c1} = (i_d^* - i_{d,vc1}^p(k+2))^2 + (i_q^* - i_{q,vc1}^p(k+2))^2 \qquad [2.44]$$

$$d_{i1} + d_{j1} \leq 1 \qquad [2.45]$$

$$d_{i1} \geq 0, \quad d_{j1} \geq 0, \qquad [2.46]$$

where $i_{v,vc1}^p(k+2)$ are the v-axis current predictions for instant $k+2$ due to the application of $\mathbf{v}_{vc,1}$ [2.43], with $v \in \{d, q\}$.

Using the Lagrange multipliers method to minimize [2.44], while ignoring the constraints [2.45]–[2.46], a relaxed solution is obtained for $\{d_{i1}, d_{j1}\}$:

$$d_{i1} = \frac{e_{d,j} \cdot e_{q,0} - e_{d,0} \cdot e_{q,j}}{e_{d,0}\left(e_{q,i} - e_{q,j}\right) + e_{d,i}\left(e_{q,j} - e_{q,0}\right) + e_{d,j}\left(e_{q,0} - e_{q,i}\right)} \qquad [2.47]$$

$$d_{j1} = \frac{e_{d,0} \cdot e_{q,i} - e_{d,i} \cdot e_{q,0}}{e_{d,0}\left(e_{q,i} - e_{q,j}\right) + e_{d,i}\left(e_{q,j} - e_{q,0}\right) + e_{d,j}\left(e_{q,0} - e_{q,i}\right)} \quad [2.48]$$

where:

$$e_{v,0} = i_v^* - i_{v,0}^p(k+2), \quad e_{v,i} = i_v^* - i_{v,vi}^p(k+2), \quad e_{v,j} = i_v^* - i_{v,vj}^p(k+2), \quad [2.49]$$

with $i_{v,0}^p(k+2)$, $i_{v,vi}^p(k+2)$ and $i_{v,vj}^p(k+2)$ being the predictions of the v-axis current for instant $k+2$ due to the application of a zero vector, $\mathbf{v}_{vi}$ and $\mathbf{v}_{vj}$, with $v \in \{d,q\}$.

However, during transients, [2.47]–[2.48] might provide unfeasible solutions for $\{d_{i1}, d_{j1}\}$, which do not satisfy the constraints [2.45]–[2.46]. In such a scenario, the modulated virtual vector is not located inside the realizable modulation region, denoted as region I in Figure 2.13. Thus, the fundamental control stage considers an additional case where the modulated virtual vector is located on the edge of the dodecagon, delimited by vectors $\mathbf{v}_{vi}$ and $\mathbf{v}_{vj}$, denoted as region II in Figure 2.13. For case II, the modulated virtual vector is defined as:

$$\mathbf{v}_{vc,2} = d_{i2} \cdot \mathbf{v}_{vi} + d_{j2} \cdot \mathbf{v}_{vj}, \quad [2.50]$$

where the duty cycles $\{d_{i2}, d_{j2}\}$ are calculated by minimizing [2.51] subjected to [2.52] and [2.53]:

$$g_{f,c2} = (i_d^* - i_{d,vc2}^p(k+2))^2 + (i_q^* - i_{q,vc2}^p(k+2))^2 \quad [2.51]$$

$$d_{i2} + d_{j2} = 1 \quad [2.52]$$

$$d_{i2} \geq 0, \quad d_{j2} \geq 0, \quad [2.53]$$

where $i_{v,vc2}^p(k+2)$ are the v-axis current predictions for instant $k+2$ due to the application of $\mathbf{v}_{vc,2}$ [2.50] with $v \in \{d,q\}$.

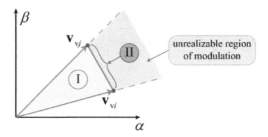

Figure 2.13. Location of the modulated virtual vector for cases I (linear region) and II (overmodulation region) considered by the fundamental stage

Considering the Lagrange multipliers method to minimize [2.51] subjected to [2.52], while ignoring [2.53], a relaxed solution is obtained for d_{i2}:

$$d_{i2} = \frac{e_{d,j}(e_{d,j} - e_{d,i}) + e_{q,j}(e_{q,j} - e_{q,i})}{(e_{d,j} - e_{d,i})^2 + (e_{q,j} - e_{q,i})^2}, \quad [2.54]$$

where d_{i2} is limited to the interval $[0,1]$. From [2.52], d_{j2} is given by:

$$d_{j2} = 1 - d_{i2} \quad [2.55]$$

In summary, the fundamental control stage determines the optimal pair of virtual vectors $\{\mathbf{v}_{vi}, \mathbf{v}_{vj}\}$ and selects $\{d_i, d_j\} = \{d_{i1}, d_{j1}\}$ if constraints in [2.52] and [2.53] are verified (case I, linear region of modulation) or $\{d_i, d_j\} = \{d_{i2}, d_{j2}\}$ otherwise (case II, overmodulation region).

Since the control of the x'-y' currents requires the application of non-zero x-y voltage components to the six-phase machine, the concept of dual virtual vectors was introduced in Gonçalves et al. (2019b). Unlike the virtual vectors, the dual virtual vectors shown in Figure 2.14 have an amplitude of $0.598 \cdot U_{dc}$ in the x-y subspace and average zero components in the α-β subspace. Hence, dual virtual vectors enable the control of the x'-y' currents without disturbing the control of the d-q currents. The dual virtual vectors are created by applying during T_s a large vector and a medium-large vector with coincident phases in the x-y subspace and the duty cycles in [2.32]. As an example, the dual virtual vector $\mathbf{v}_{dvl}$ results from the application of the large vector $\mathbf{v}_{34}$ during $d_L \cdot T_s$ and the medium-large vector $\mathbf{v}_{43}$ during $d_{ML} \cdot T_s$:

$$\begin{aligned}\mathbf{v}_{dvl} &= \mathbf{v}_{34}\, d_L + \mathbf{v}_{43}\, d_{ML} \\ &= U_{dc} \begin{bmatrix} 0 & 0 & \frac{\sqrt{3}}{3} & \frac{2\sqrt{3}}{3} - 1 \end{bmatrix}^T \cdot\end{aligned} \quad [2.56]$$

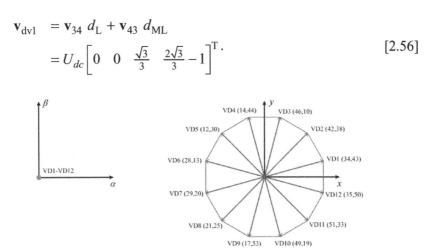

Figure 2.14. *Mapping of the dual virtual vectors into the (a) $\alpha-\beta$ subspace and (b) x-y subspace*

The secondary control stage is responsible for regulating the x'-y' currents and selecting two optimal dual virtual vectors $\{\mathbf{v}_{dvm}, \mathbf{v}_{dvn}\}$ along with their respective duty cycles $\{d_m, d_n\}$. The first step of the fundamental control stage is to predict the x'-y' currents for instant $k+1$ with:

$$i_{x'}^p(k+1) = \left(1 - \frac{R_s T_s}{L_{dq}}\right) i_{x'}(k) - \omega_r(k) T_s i_{y'}(k) + \frac{T_s}{L_{xy}}\left(u_{x'}(k) - e_{x'}(k)\right) \quad [2.57]$$

$$i_{y'}^p(k+1) = \left(1 - \frac{R_s T_s}{L_{dq}}\right) i_{y'}(k) + \omega_r(k) T_s i_{x'}(k) + \frac{T_s}{L_{xy}}\left(u_{y'}(k) - e_{y'}(k)\right), \quad [2.58]$$

where, according to [2.29], the x'-y' back-EMF components considering up to the seventh-order harmonic, are given by:

$$e_{x'}(k) = -5\omega_r(k)\psi_{PM5}\sin(6\,\theta(k)+\phi_5) - 7\omega_r(k)\psi_{PM7}\sin(6\,\theta(k)+\phi_7) \quad [2.59]$$

$$e_{y'}(k) = 5\omega_r(k)\psi_{PM5}\cos(6\theta(k)+\phi_5) - 7\omega_r(k)\psi_{PM7}\cos(6\theta(k)+\phi_7). \quad [2.60]$$

The predictions of the *d-q* currents for instant $k+2$, due to the application of an *c*-index virtual vector $\mathbf{v}_{vc}$, are calculated by:

$$i^p_{x',da}(k+2) = \left(1 - \frac{R_s T_s}{L_{xy}}\right)i^p_{x'}(k+1) - \omega_r(k)\, T_s\, i^p_{y'}(k+1) + \frac{T_s}{L_{xy}}\left(u_{x'}(k+1) - e_{x'}(k+1)\right) \quad [2.61]$$

$$i^p_{y',da}(k+2) = \left(1 - \frac{R_s T_s}{L_{xy}}\right)i^p_{y'}(k+1) + \omega_r(k)\, T_s\, i^p_{x'}(k+1) + \frac{T_s}{L_{xy}}\left(u_{y'}(k+1) - e_{y'}(k+1)\right) \quad [2.62]$$

where the *x-y* voltages for instant $k+1$ are given by:

$$\begin{bmatrix} 0 & 0 & u_{x'}(k+1) & u_{y'}(k+1) \end{bmatrix}^T = \mathbf{R}\cdot\mathbf{v}_{da}, \quad a \in \{1,...,12\}, \quad [2.63]$$

and the back-EMF components are calculated using [2.29], considering up to the seventh harmonic:

$$\begin{aligned} e_{x'}(k+1) = &-5\omega_r(k)\,\psi_{PM5}\sin(6\theta(k+1)+\phi_5) \\ &-7\omega_r(k)\,\psi_{PM7}\sin(6\theta(k+1)+\phi_7) \end{aligned} \quad [2.64]$$

$$\begin{aligned} e_{y'}(k+1) = &5\omega_r(k)\,\psi_{PM5}\cos(6\theta(k+1)+\phi_5) \\ &-7\omega_r(k)\,\psi_{PM7}\cos(6\theta(k+1)+\phi_7) \end{aligned}, \quad [2.65]$$

where $\theta(k+1)$ is the rotor electrical position for instant $k+1$, given by:

$$\theta(k+1) = \theta(k) + T_s\,\omega_r(k). \quad [2.66]$$

The secondary control stage determines the indexes $\{m,n\}$ of the optimal pairs of virtual vectors $\{\mathbf{v}_{dvm}, \mathbf{v}_{vdn}\}$ with:

$$\{m,n\} = \arg\min_{\{a,b\}}(g_{s,a} + g_{s,b}), \quad \{a,b\} = \{\{1,2\},\{2,3\},...,\{12,1\}\}, \quad [2.67]$$

where $\{g_{s,a}, g_{s,b}\}$ are the cost functions due to the application of $\{\mathbf{v}_{dvm}, \mathbf{v}_{vdn}\}$, which are neighbours in the x-y subspace (Figure 2.14), and are given by:

$$g_{s,a} = \left(i_{x'}^* - i_{x',da}^p(k+2)\right)^2 + \left(i_{y'}^* - i_{y',da}^p(k+2)\right)^2 \qquad [2.68]$$

$$g_{s,b} = \left(i_{x'}^* - i_{x',db}^p(k+2)\right)^2 + \left(i_{y'}^* - i_{y',db}^p(k+2)\right)^2 \qquad [2.69]$$

where the x'-y' current predictions are calculated with [2.61]–[2.62] for the candidate dual virtual vectors $\{\mathbf{v}_{dvm}, \mathbf{v}_{vdn}\}$.

After determining with [2.67] the indexes $\{m,n\}$ for the optimal pair of dual virtual vectors, the secondary control stage calculates the optimal duty cycles of these vectors $\{d_m, d_n\}$ to minimize the tracking error of the x'-y' currents. By defining a dual modulated virtual vector as:

$$\mathbf{v}_{dc,1} = d_{m1} \cdot \mathbf{v}_{dm} + d_{n2} \cdot \mathbf{v}_{dn}, \qquad [2.70]$$

with $\mathbf{v}_{dc,1}$ located inside the dodecagon of Figure 2.9, the duty cycles $\{d_{m1}, d_{n1}\}$ are calculated by minimizing [2.71], subjected to [2.72]–[2.73]:

$$g_{s,c1} = (i_{x'}^* - i_{x',dc1}^p(k+2))^2 + (i_{y'}^* - i_{y',dc1}^p(k+2))^2 \qquad [2.71]$$

$$d_{m1} + d_{n1} \leq d_{xy}^{max} \qquad [2.72]$$

$$d_{m1} \geq 0, \quad d_{n1} \geq 0, \qquad [2.73]$$

where in order to account for the output voltage limit of the 2L-VSIs, d_{xy}^{max} is calculated by:

$$d_{xy}^{max} = 1 - d_i - d_j \qquad [2.74]$$

Using the method of Lagrange multipliers to minimize [2.71] while discarding constraints [2.72]–[2.73], we obtain the following relaxed solution:

$$d_{m1} = \frac{e_{x',n} \cdot e_{y',0} - e_{x',0} \cdot e_{y',n}}{e_{x',0}\left(e_{y',m} - e_{y',n}\right) + e_{x',m}\left(e_{y',n} - e_{y',0}\right) + e_{x',n}\left(e_{y',0} - e_{y',m}\right)} \qquad [2.75]$$

$$d_{n1} = \frac{e_{x',0} \cdot e_{y',m} - e_{x',m} \cdot e_{y',0}}{e_{x',0}\left(e_{y',m} - e_{y',n}\right) + e_{x',m}\left(e_{y',n} - e_{y',0}\right) + e_{x',n}\left(e_{y',0} - e_{y',m}\right)}, \qquad [2.76]$$

with:

$$e_{v,0} = i_v^* - i_{v,0}^p(k+2), \quad e_{v,m} = i_v^* - i_{v,dm}^p(k+2), \quad e_{v,n} = i_v^* - i_{v,dn}^p(k+2), \qquad [2.77]$$

where $i_{v,dm}^p(k+2)$ and $i_{v,dn}^p(k+2)$ are the predictions of the v-axis current to instant $k+2$ due to the application of $\mathbf{v}_{dm}$ and $\mathbf{v}_{dn}$, with $v \in \{x',y'\}$.

However, [2.75]–[2.76] might provide unfeasible solutions during transients, which do not satisfy the constraints in [2.72]–[2.73]. In that case, the secondary stage considers a second case (case II), similar to the fundamental stage, where the modulated dual virtual vector is defined as:

$$\mathbf{v}_{dc,2} = d_{m2} \cdot \mathbf{v}_{dm} + d_{n2} \cdot \mathbf{v}_{dn}, \qquad [2.78]$$

where the optimal duty cycles minimize [2.79] subjected to [2.80] and [2.81]:

$$g_{s,c2} = (i_{x'}^* - i_{x',dc2}^p(k+2))^2 + (i_{y'}^* - i_{y',dc2}^p(k+2))^2 \qquad [2.79]$$

$$d_{m2} + d_{n2} = d_{xy}^{max} \qquad [2.80]$$

$$d_{m2} \geq 0, \quad d_{n2} \geq 0. \qquad [2.81]$$

Control and Diagnosis of Faults 65

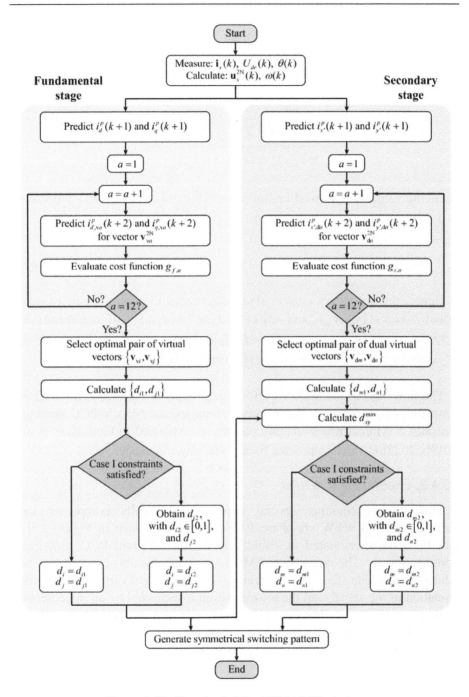

Figure 2.15. *Flowchart of the MSVV-PCC strategy*

An unconstrained solution for the duty cycle d_{m2} is given by:

$$d_{m2} = \frac{\left(i_{x'}^* - d_{xy}^{\max} i_{x',dn}^p(k+2)\right)\left(e_{x',m} - e_{x',n}\right)^2}{\left(e_{x',m} - e_{x',n}\right)^2 + \left(e_{y',m} - e_{y',n}\right)^2} + \frac{\left(i_{y'}^* - d_{xy}^{\max} i_{y',dn}^p(k+2)\right)\left(e_{y',m} - e_{y',n}\right)^2}{\left(e_{x',m} - e_{x',n}\right)^2 + \left(e_{y',m} - e_{y',n}\right)^2} \quad [2.82]$$

where the value of d_{m2} must be limited to $\left[0, d_{xy}^{\max}\right]$. From [2.80], the value of d_{n2} is calculated by:

$$d_{n2} = d_{xy}^{\max} - d_{m2}. \quad [2.83]$$

Hence, the secondary control stage determines the optimal pair of dual virtual vectors $\{\mathbf{v}_{dm}, \mathbf{v}_{dn}\}$ and selects $\{d_m, d_n\} = \{d_{m1}, d_{n1}\}$ if constraints in [2.72]–[2.73] are verified (case I, linear region of modulation) or $\{d_m, d_n\} = \{d_{m2}, d_{n2}\}$ otherwise (case II, overmodulation region).

The flowchart of the MSVV-PCC strategy is shown in Figure 2.15. After completing the execution of both control stages, the MSVV-PCC strategy generates a symmetrical switching pattern, as explained in Gonçalves et al. (2019b, 2022b), which imposes a fixed switching frequency $\bar{f}_{sw} = 1/T_s$.

2.5.4.2. Experimental results

This section contains several experimental results comparing the performance of a 4 kW six-phase PMSG drive, as shown in Figure 2.16, with the parameters listed in Table 2.1, for the different PCC strategies aforementioned. The six-phase PMSG is mechanically coupled to an induction machine fed by a variable frequency converter to provide regulation of the speed, and the rotor position is measured by an incremental encoder with 2048 ppr. The control strategies are implemented in a dSPACE dS1103 platform and the symmetrical switching patterns are generated by the FPGA of a National Instruments cRIO-9066.

Parameter	Value	Parameter	Value	Parameter	Value
P_s (kW)	4	R_s (Ω)	1.5	ϕ_5 (deg)	1.3
U_s (V)	340	L_{dq} (mH)	53.8	ϕ_7 (deg)	−12.7
I_s (A)	3.4	L_{xy} (mH)	2.1	T_s (μs)	35, 50, 100
n_n (rpm)	1,500	$\psi_{s,PM1}$ (mWb)	980.4	U_{dc} (V)	650
T_n (N·m)	28.2	$\psi_{s,PM5}$ (mWb)	2.4	T_d (μs)	2.2
p	2	$\psi_{s,PM7}$ (mWb)	1.6		

Table 2.1. *Parameters of the experimental setup*

Figure 2.16. *Experimental setup*

The experimental results for the steady-state operation of the six-phase PMSG drive (generating mode) at rated speed and load conditions are shown in Figure 2.17. The current spectra of phase a_1, obtained for all control strategies, are shown in Figure 2.18. To evaluate the harmonic content of currents, including the ripple, the total harmonic distortion (THD) index is defined as:

$$\text{TWD}_i = \frac{1}{6} \sum_{\chi=a1,\ldots,c2} \frac{\sqrt{I_\chi^2 - i_{\chi,1}^2}}{i_{\chi,1}} \times 100\%, \quad [2.84]$$

where I_χ is the root mean square (RMS) value of current in phase χ, which is calculated as:

$$I_\chi = \sqrt{\frac{1}{N} \sum_{n=1}^{N} i_\chi^2(n)}, \quad \chi \in \{a1, b1, c1, a2, b2, c2\}. \quad [2.85]$$

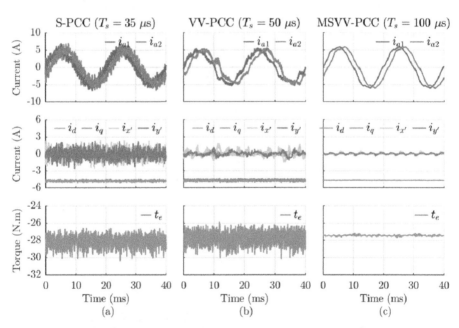

Figure 2.17. Experimental results for the PMSG drive operating at 100% of the rated speed and load for (a) S-PCC, (b) VV-PCC and (c) MSVV-PCC

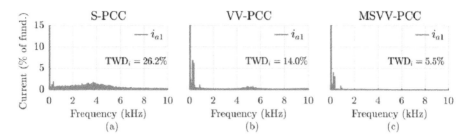

Figure 2.18. *Experimental results for the current spectra in phase a1, considering the PMSG drive operating at 100% of the rated speed and load for (a) S-PCC, (b) VV-PCC and (c) MSVV-PCC*

From the inspection of Figures 2.17 and 2.18, the S-PCC strategy gives the worst result among the considered strategies in terms of current distortion (TWD_i) and current ripple in the x'-y' subspace, which should be minimized to reduce the copper losses in the PMSG. On the other hand, the VV-PCC strategy significantly reduces the current distortion due to the use of virtual vectors, which inject zero average voltage components in the x-y subspace. However, since the VV-PCC strategy cannot regulate the currents in the x'-y' subspace, non-negligible fifth- and seventh-order harmonics appear in the phase currents, which are mainly caused by the dead-time effects in the inverters and back-EMF harmonics. The use of modulated virtual vectors by the MSVV-PCC strategy significantly reduces the current distortion, reducing the value of the TWD_i to only 5.5%. In terms of switching frequency, the S-PCC and VV-PCC strategies provide an average switching frequency of 3.92 kHz and 11.11 kHz, respectively, while the MSVV-PCC gives a fixed switching frequency of 10 kHz.

The experimental results for the operation of the six-phase PMSG drive at 1,200 rpm with a 50% load level are shown in Figure 2.19. The secondary control stage is disabled in the case of Figure 2.19(a), whereas it is enabled in the case of Figure 2.19(b). Furthermore, the corresponding current spectra in phase a_1 are presented in Figure 2.20 for the mentioned scenarios. From the inspection of these results, the activation of the secondary control stage suppresses the x-y currents and reduces the TWD_i from 17.5% to 6.1%.

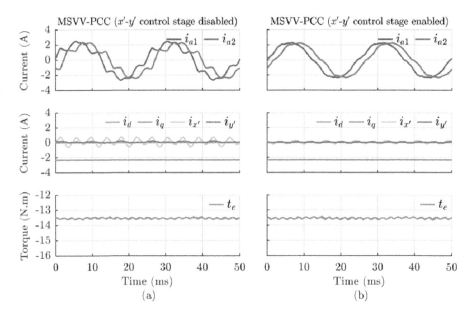

Figure 2.19. Experimental results for the PMSM drive operating at 1,200 rpm with 50% load, under the MSVV-PCC strategy: (a) secondary (x'-y') control stage disabled; (b) secondary (x'-y') control stage enabled

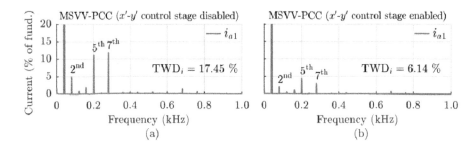

Figure 2.20. Experimental results for the a1-phase current spectrum, considering the PMSM drive operating at 1,500 rpm with 50% of the rated load (generating mode) under the MSVV-PCC strategy with (a) the secondary (x'-y') control stage disabled and (b) the secondary (x'-y') control stage disabled

2.6. Fault diagnosis in multiphase drives

2.6.1. *Introduction*

Although multiphase machines and drives have inherent fault-tolerant capabilities, it is still essential to quickly detect, identify and isolate any faults that may arise. This will allow the system to continue operating in a fault-tolerant mode until scheduled maintenance can be performed. These features are essential for offshore WECS, where repair and maintenance actions are both difficult and expensive.

Like in the case of three-phase drives, numerous types of faults may appear in multiphase drives. Figure 2.21 presents the main types of faults in multiphase PMSG drives.

Although no reliability studies have been conducted so far specifically for multiphase drives, the failure rates of their different components are expected to be similar to those found in their three-phase counterparts. Therefore, regarding the power electronic converters, IGBT faults like open- and short-circuit faults are expected to represent a significant portion of faults in this drive component. Regarding the electric machine, insulation failures in the stator windings are expected to be the dominant cause of stator faults, particularly in the form of interturn short-circuits. Regarding the rotor of the machine, PM demagnetization faults is the dominant type of fault. Sensor faults, either in the machine (encoder) or in the power converters (current sensors and voltage sensors measuring the DC-link and grid voltages), should not be neglected as they can also negatively impact the operation of the drive. Naturally, faults of mechanical nature, such as those in the machine bearings and eventual gearboxes, should also be considered, as in the case of three-phase drives.

In the following sections, some of these types of faults will be addressed in more detail, focusing on some diagnostic methods available to diagnose them.

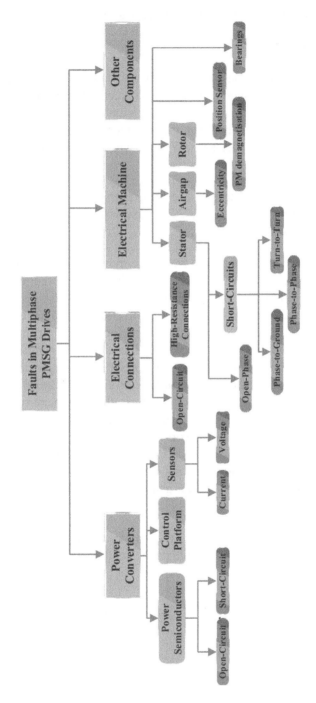

Figure 2.21. Classification of faults in multiphase PMSG drives

2.6.2. Interturn short-circuit faults

Stator winding faults are one of the most prevalent types of electrical faults in electric machines. Their causes are diverse and usually result from the combination of several electrical, mechanical, thermal and environmental stresses acting on the machine. These faults are typically categorized as interturn (or turn-to-turn), phase-to-phase and phase-to-ground faults.

Interturn short-circuits (ITSCs), especially the ones involving only one or a few adjacent turns of the same phase, are the most troublesome kind of stator faults and, consequently, are often addressed in the specialized literature. Their rapid detection and isolation are critical to avoid further damage and a catastrophic failure of the machine, particularly in PMSMs where the back-EMF is non-zero even after the disconnection of the inverter(s) supplying the stator windings, thus continuing to impose the circulation of a short-circuit current.

The detection of ITSCs in a short period of time is particularly important because the short-circuit current in the affected turns can be very high. This can lead to a significant temperature rise and cause further damage to the insulation system of the remaining winding. In the case of PMSMs, it may even result in eventual PM demagnetization.

Some diagnostic techniques proposed in the literature for three-phase machines can be readily adapted for multiphase machines, while others, especially developed for multiphase machines, take advantage of the specificities of this type of machine, mainly related to the existence of different subspaces in their mathematical model (Immovilli et al. 2015; Faiz et al. 2017).

A short-circuit in one of the stator windings which disrupts the symmetry of the machine and leads to the loss of independence between the different subspaces that characterise its normal operation. This spreads the effects of a stator fault across the different subspaces. An ITSC leads to a short-circuit current that creates a negative sequence component in the voltage and/or current components mapped into the stationary fundamental subspace ($\alpha-\beta$ subspace). The effects of the ITSC may be more pronounced in the current or voltage signals, depending on the bandwidth of the current controllers. For the typical high bandwidths of the current control loops in field-oriented control schemes, and for PCC schemes, the effects of the fault will be mostly visible in the voltage signals. The voltage space vector in the $\alpha-\beta$ subspace will thus contain a negative sequence voltage component at a frequency of $-f_1$, in addition to the fundamental component at a frequency f_1. The ratio between these two components gives an indication of the extension of the fault, which is considered proportional to the product between the short-circuit current and the number of shorted turns, as is usually done in the diagnosis of ITSC in three-phase machines. The fault will also manifest itself in the x-y subspace, although the effects in this subspace seem to be dependent on the type of winding configuration of the machine. Finite element simulations performed with an asymmetrical six-phase PMSG drive with distributed windings show that the components at the frequencies f_1 and $-f_1$ appear in the spectrum of the voltage space vector in the x-y subspace due to an ITSC (Figure 2.22). Meanwhile, in Bianchini et al. (2008), using a five-phase PMSM with concentrated windings, it was found that the voltage component at the frequency $-3f_1$ is the best indicator of ITSCs in the x-y subspace. A subsequent work, using the same type of machine, proposed a novel approach to diagnose ITSC, whose philosophy relies on the combined used of the information obtained in two subspaces (Immovilli et al. 2015). To avoid the loss of information and to maximize the variations in the spectral content between the healthy and the ITSC faulty case, it was proposed to use the product of the voltage space vectors in the $\alpha-\beta$ and x-y subspaces ($\underline{u}_{\alpha\beta}$ and $\underline{u}_{xy}$, respectively):

$$\underline{D} = \underline{u}_{\alpha\beta} \times \underline{u}_{xy} \qquad [2.86]$$

The fault can be detected by the analysis of the spectrum of $\underline{D}$ by monitoring the amplitudes of the DC and $2f_1$ components, which increase with the severity of the ITSC fault.

Similar to the case of three-phase machines, we can also use the zero-sequence voltage component to detect ITSCs in multiphase machines. Under ideal conditions, in a healthy and symmetrical machine, the algebraic sum of the fundamental component of each phase voltage is zero and the zero-sequence voltage component (ZSVC) will only contain harmonic components at the frequency $3f_1$ and odd multiples of it (all created by the PM flux density harmonics). With the appearance of an ITSC, the ZSVC will contain a new component at the fundamental frequency f_1, being the main indicator of the fault. However, the implementation of this method requires access to the neutral point of the machine and the installation of as many additional voltage sensors as the number of phases, making its implementation costly. A cheaper and simpler alternative for six-phase machines with two isolated neutral points consists of, instead, monitoring the voltage between the two neutral points. The harmonic content of this voltage will include a component at the fundamental frequency f_1 when there are ITSCs. The amplitude of this component increases consistently with the extension of the fault. The main drawback of this approach is the difficulty/impossibility of locating the faulty phase (Armindo 2021).

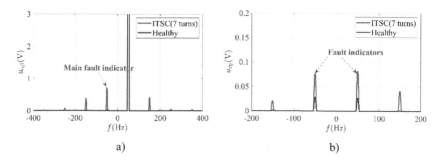

Figure 2.22. Effects of an ITSC fault in an asymmetrical six-phase PMSG: (a) primary subspace; (b) secondary subspace

The spectrum analysis of the active and reactive powers of the machine can also be used to diagnose ITSCs in multiphase machines. Using this method, the ITSC fault leads to the appearance of an additional component in such spectra, which appears at $2f_1$. Additionally, Wang et al. (2018) have shown that the reactive power is more sensitive to the fault when the machine operates as a motor, while the active power is more sensitive to the fault with the machine in generating mode.

Other techniques have been proposed for the detection of ITSCs in multiphase machines, especially in fault-tolerant machines with concentrated windings. One such technique consists of using the PWM ripple current as in Hu et al. (2017), but its implementation requires additional hardware. Another method, proposed for a five-phase machine with open-end windings, is based on the analysis of the zero-sequence current component (Fan et al. 2017), but its extension is not straightforward for multiphase machines with other winding connections, which are usually used in WECS.

For a more comprehensive list of other diagnostic techniques, the interested reader is referred to Yepes et al. (2022b).

2.6.3. *High-resistance connections and open-phase faults*

HRCs are a type of fault in electrical machines that may result from damaged power connections in the stator terminals due to corrosion, vibration or unsecured connections. In a worst-case scenario, HRCs can evolve into an OPF (Gonçalves et al. 2022c). On the other hand, OPFs are commonly addressed in the literature and can occur either as an open insulated-gate bipolar transistor (IGBT) fault or an open-circuit in one of the motor phases or phase connections between the converters and the machine (Munim et al. 2017). It must be highlighted that an OPF may be the result of completely opening a phase (e.g. using fuses or bidirectional switches) as a solution to many other kinds of faults, such as IGBT open- and short-circuit faults, HRCs, current-sensor faults, or the isolation of a converter module (Yepes et al. 2022a). Thus, several other types of faults, namely IGBT short-circuit faults, ultimately lead to OPFs. This justifies the main focus on OPFs in this chapter.

Several methods have been proposed to diagnose OPFs and HRCs in multiphase machines. The available methods for the detection of OPFs can be broadly classified into model-based, knowledge-based and signal-based strategies (Guo et al. 2021). Model-based methods have been widely used to detect faults in drive systems. By employing different types of observers, such as the Luenberger observer, sliding mode observer, nonlinear observers or variable structure observers, this category of methods is effective in the diagnosis of faults as they have relatively short detection times. In addition, model-based strategies do not require extra hardware and can be insensitive to load changes. However, since, in general, these methods are based on a residual generation or some correlation between measured and estimated quantities, their performance is highly dependent on the accuracy of the mathematical model used and sensitive to variations in the machine parameters. As an example, in the model-based strategy presented by Salehifar et al. (2015) for a five-phase BLDC motor, a model-based observer is designed to estimate the model parameters. In conjunction with a sliding mode observer, these parameters are then used to estimate the phase currents using an ideal motor model. The similarity between the estimated and real currents is subsequently measured using a cross-correlation factor. This factor is finally used to define a fault detection index. This diagnostic scheme is simple and fast, being able to detect multiple open switch and OPFs in the machine.

Knowledge-based diagnostic methods for power switch fault diagnosis are based on advanced algorithms such as expert systems, fuzzy logic, neural networks or deep learning techniques. Since these methods are based on system historical data, they do not require an accurate system model. However, these methods usually need a long training time, leading to a significant computational effort, and significantly increasing their implementation difficulty. These limitations make these techniques difficult to be integrated into the drive controller and too complex for real-time implementation (Guo et al. 2021; Liu et al. 2022).

Signal-based strategies, on the other hand, are by far the more common category of methods proposed in the literature to diagnose OPFs. Usually, these methods rely on the analysis of current or voltage signals, but the theory behind them is vast, as their operating philosophy is quite diverse. For instance, Arafat et al. (2017) proposed monitoring the symmetrical components of currents to detect single, double adjacent and double non-adjacent OPFs in five-phase PMSM drives. Alternatively, the method in Trabelsi et al. (2016) uses a centroid-based approach to monitor the geometric shape formed by the measured x-y currents, combined with a normalisation procedure for the currents, to identify OPFs. A different approach is proposed in Chen et al. (2021), where the second harmonic of the magnetic field pendulous oscillation position is monitored to detect changes in the magnetic field of the machine caused by OPFs. Many of these methods are based on the analysis of the machine supply currents in the fundamental and secondary subspaces. Some of them are only applicable to steady-state conditions, while others assume that the x-y currents are not directly controlled by the control system of the drive, which is a limitation in PMSG drives where such quantities need to be controlled, as shown before. A detailed analysis of all these techniques is beyond the scope of the present chapter. The interested reader is referred to Yepes et al. (2022a) where a detailed analysis and comparison of the available methods can be found.

Regarding HRCs, their detection is also important, as highlighted before. If possible, the same method should be able to detect and discriminate OPFs and HRCs. Two recent methods with such capabilities are presented in Gonçalves et al. (2022c) for asymmetrical six-phase PMSMs. Using a model predictive current-controlled drive, these faults are detected and identified using the reference and prediction current errors in the x'-y' subspace. OPFs and HRCs can be detected and quantified in one electric period, being robust to transients and parameter mismatch errors.

Many methods are proposed in the literature for detecting HRCs, including those that analyse currents and voltages in the primary and secondary subspaces of multiphase machines. Interested readers can find a comprehensive list of such methods in Yepes et al. (2022b).

2.6.4. *Permanent magnet faults*

The demagnetisation of the PMs in the rotor of PMSMs may occur for several reasons. The most common causes of PM demagnetization are thermal stresses due to large currents in the stator windings, deficiencies in the cooling system of the machine, mechanical stresses, cracks and high demagnetization fields due to negative d-axis currents created by stator faults such as interturn short-circuits, manufacturing defects and corrosion/oxidation of the PMs.

There are two types of PM demagnetization: (a) uniform demagnetisation and (b) partial/local demagnetisation. Uniform demagnetization affects all rotor magnets equally, while local demagnetization affects only some of the magnets.

An immediate consequence of uniform demagnetisation is the reduction of the amplitude of the back-EMF waveform, leading to a reduction of the torque developed by the machine for the same value of the stator current (Yepes et al. 2022b). In this type of PM fault, no new harmonics are expected in the machine line currents, supply voltages or back-EMF. On the other hand, in the case of local demagnetization, in addition to reducing the amplitude of the back-EMF, their high-order harmonics will suffer some changes. In particular, the fifth- and seventh-order harmonics, mapped into the x-y subspace, change to some extent their amplitude with the degree of demagnetisation (Gritli et al. 2020). If these back-EMF harmonics are minimized during the design stage of the machine, their amplitude will suffer an increase (Gritli et al. 2019). On the other hand, if such minimization is not done at the design stage, those harmonics are still expected to change their amplitude and/or phase, possibly detecting the fault if they are adequately characterized during the regular operation of the machine. Once present in the back-EMF, these harmonics will also appear in the supply voltages in the x-y subspace if there is a closed-loop current control system that imposes constant values, usually null, for the reference currents in such subspace, which is a commonly adopted measure to improve the machine efficiency and decrease the THD of the machine supply currents. In general, depending on the bandwidth of the current control loops, the effects of the fault may be more pronounced in the voltage signals and current signals or be distributed between these two signals. If the control system does not regulate the x-y current components, the effects of the PM demagnetization will be more visible in the current harmonics, mapped into the x-y subspace, created by the corresponding back-EMF harmonics combined with the low impedance of the machine in this subspace.

We should bear in mind that the fifth and the seventh harmonics are also present in the currents and/or supply voltages of the machine in the x-y subspace due to the inverter dead-time (Gonçalves et al. 2019b, 2022a). Thus, a proper characterization of the amplitude and phase of such harmonics with the machine in healthy conditions is always advisable.

In the case of multiphase machines with a number of phases multiple of three, there are other options for detecting PM magnetisation faults. In the case of a six-phase PMSM with two isolated neutral points, a possibility is to monitor the voltage between the two neutral points, being the fault detected by the observation of a significant reduction in the amplitude of the third harmonic that exists in such voltage (Figueiras 2021).

2.6.5. Current sensor faults

Current sensor faults are another type of fault that has a negative impact on the operation of the drive. Current sensor faults are usually classified as noise in the output, zero output, gain error and DC offset.

In Wang et al. (2019), it is shown that current sensor faults (zero output) in an asymmetrical six-phase PMSM drive with uncontrolled x-y currents lead to current trajectories in the x-y subspace, obtained with the measured currents, which are straight lines whose slope is associated with the phase where the faulty current sensor is located. It should be noted that these straight lines appear, although the real machine currents in the x-y subspace are still very small even with current sensor faults, demonstrating the resilience of multiphase drives to this type of fault. For OPFs, straight lines are also obtained in the same x-y subspace but with different slopes, thus allowing the discrimination of the two types of faults. However, if the machine currents in this subspace are controlled, or if the control system includes an observer to compensate for disturbances and parameter mismatch errors, the behavior of the drive will differ, thus invalidating the conclusions above.

In Yao et al. (2020), a different diagnostic method was proposed for identifying OPFs and current sensor faults. The authors propose to perform fault detection and feature extraction for OPFs and current sensor faults by analyzing the d-axis component of the generator output currents using empirical mode decomposition, the Hilbert transform, variable-parameter particle swarm optimization and a support vector machine. They considered a five-phase PMSG in their study. While the presented simulation results seem to validate the use of this diagnostic methodology, it is a complex and time-consuming procedure to detect faults using this approach. Furthermore, a diode-bridge rectifier was connected to the generator in that study, being uncertain the behavior of the diagnostic system if an active rectifier was used instead.

2.6.6. *Speed sensor faults*

AC drives used in WECS rely on position sensors, such as encoders or resolvers, that may be the source of faults, thus compromising the electrical and mechanical integrity of the entire WECS. Typical faults in incremental encoders are an inconsistent number of pulses per rotor turn (e.g. due to noise, internal problems in the encoder itself or in the cable connections between the encoder and control board) and intermittent or total loss of one or more output signals (Wang et al. 2019).

A standard method for detecting encoder/resolver faults is based on the comparison of the speed measurement with an estimated speed value obtained for instance with the aid of observers of different types (Salmasi 2017; Bensalem et al. 2021). If the difference between those two values is higher than a given threshold, a fault is signaled. The problem with this diagnostic approach is that the difference between the measured and estimated speed values can arise as well due to other types of machine or converter faults. A possible solution to circumvent this problem is the use of more than one observer, based on different variables, to obtain the estimated speed, or the combined use of other diagnostic algorithms to detect the other types of faults concurrently. Alternatively, another approach to overcome this difficulty is the implementation of a diagnosis test using the d- and q-reference current components, proposed in Salmasi (2017) for three-phase machines but also applicable for the case of multiphase machines.

2.7. Conclusion

This chapter has highlighted the importance of multiphase PMSG drives, particularly the ones used in high-power WECS operating in offshore locations. The mathematical modeling of multiphase PMSGs is addressed and, as a case study, the detailed model of an asymmetrical six-phase PMSG is presented. This model is first developed in the natural reference frame and then transformed using the widely known VSD transformation.

Different control strategies, including FOC, DTC and MPC, are then briefly discussed and reviewed for multiphase PMSGs. Greater emphasis and deeper discussion are provided for more up-to-date control algorithms based on FCS-MPC, especially those taking advantage of virtual voltage vectors. Several FCS-MPC methods are presented and compared side-by-side, with the goal of demonstrating the need for and availability of techniques for controlling the machine in both the primary and secondary subspace in order to decrease the current THD and increase the machine efficiency.

Afterward, a general overview is presented about the types of faults that may arise in multiphase PM drives, along with their causes. Different diagnostic techniques for multiphase drives are then briefly reviewed and analyzed to highlight their advantages and limitations. It is noted that these diagnostic techniques are closely related to the type of machine and control strategy being used. Through this discussion and the presentation of some results, it is demonstrated that further research is needed in this domain, to cover the diagnosis of faults in multiphase drives with different designs and control strategies. The final goal is to develop effective techniques for diagnosing faults in any multiphase drive, which will lead to improved performance, reliability and safety of these machines.

2.8. References

Abad, G., Lopez, J., Rodriguez, M., Marroyo, L., Iwanski, G. (2011). *Doubly Fed Induction Machine: Modeling and Control for Wind Energy Generation*, 1st edition. John Wiley & Sons, New York.

Anaya-Lara, O., Campos-Gaona, D., Moreno-Goytia, E., Adam, G. (2014). *Offshore Wind Energy Generation: Control, Protection, and Integration to Electrical Systems*, 1st edition. John Wiley & Sons, New York.

Anaya-Lara, O., Tande, J.O., Uhlen, K., Merz, K. (2018). *Offshore Wind Energy Technology*, 1st edition. John Wiley & Sons, New York.

Arafat, A., Choi, S., Baek, J. (2017). Open-phase fault detection of a five-phase permanent magnet assisted synchronous reluctance motor based on symmetrical components theory. *IEEE Transactions on Industrial Electronics*, 64(8), 6465–6474.

Armindo, V.A. (2021). Analysis and diagnosis of faults in the stator windings of six-phase predictive current controlled PMSMs. M.Sc. Thesis, University of Coimbra.

Bensalem, Y., Abbassi, R., Jerbi, H. (2021). Fuzzy logic based-active fault tolerant control of speed sensor failure for five-phase PMSM. *Journal of Electrical Engineering & Technology*, 16(1), 287–299.

Bianchini, C., Fornasiero, E., Matzen, T.N., Bianchi, N., Bellini, A. (2008). Fault detection of a five-phase permanent-magnet machine. In *34th Annual Conference of IEEE Industrial Electronics*, Orlando.

Blaabjerg, F. and Ma, K. (2013). Future on power electronics for wind turbine systems. *IEEE Journal of Emerging and Selected Topics in Power Electronics*, 1(3), 139–152.

Cardenas, R., Peña, R., Alepuz, S., Asher, G. (2013). Overview of control systems for the operation of DFIGs in wind energy applications. *IEEE Transactions on Industrial Electronics*, 60(7), 2776–2798.

Chen, H., He, J., Demerdash, N.A.O., Guan, X., Lee, C.H.T. (2021). Diagnosis of open-phase faults for a five-phase PMSM fed by a closed-loop vector-controlled drive based on magnetic field pendulous oscillation technique. *IEEE Transactions on Industrial Electronics*, 68(7), 5582–5593.

Cordovil, P.T.C. (2018). Modelagem e Aspectos Construtivos de Geradores Síncronos Multifásicos para Turbinas Eólicas Offshore. Master's Thesis, University of São Paulo.

Duran, M.J., Levi, E., Barrero, F. (2017). Multiphase electric drives: Introduction. In *Wiley Encyclopedia of Electrical and Electronics Engineering*. John Wiley & Sons, New York.

Eldeeb, H.M., Abdel-Khalik, A.S., Hackl, C.M. (2019). Dynamic modeling of dual three-phase IPMSM drives with different neutral configurations. *IEEE Transactions on Industrial Electronics*, 66(1), 141–151.

Faiz, J., Nejadi-Koti, H., Valipour, Z. (2017). Comprehensive review on inter-turn fault indexes in permanent magnet motors. *IET Electric Power Applications*, 11(1), 142–156.

Fan, Y., Li, C., Zhu, W., Zhang, X., Zhang, L., Cheng, M. (2017). Stator winding interturn short-circuit faults severity detection controlled by OW-SVPWM without CMV of a five-phase FTFSCW-IPM. *IEEE Transactions on Industry Applications*, 53(1), 194–202.

Faulstich, S., Hahn, B., Tavner, P.J. (2011). Wind turbine downtime and its importance for offshore deployment. *Wind Energy*, 14(3), 327–337.

Feng, G., Lai, C., Kelly, M., Kar, N.C. (2019). Dual three-phase PMSM torque modeling and maximum torque per peak current control through optimized harmonic current injection. *IEEE Transactions on Industrial Electronics*, 66(5), 3356–3368.

Figueiras, H.S. (2021). Diagnosis of permanent magnet faults in six-phase PMSMs. M.Sc. Thesis, University of Coimbra.

Geyer, T. (2016). *Model Predictive Control of High Power Converters and Industrial Drives*, 1st edition. John Wiley & Sons, New York.

Gonçalves, P.F.C., Cruz, S.M.A., Mendes, A.M.S. (2019a). Finite control set model predictive control of six-phase asymmetrical machines – An overview. *Energies*, 12(24), 4693.

Gonçalves, P.F.C., Cruz, S.M.A., Mendes, A.M.S. (2019b). Bi-subspace predictive current control of six-phase PMSM drives based on virtual vectors with optimal amplitude. *IET Electric Power Applications*, 13(11), 1672–1683.

Gonçalves, P.F.C., Cruz, S.M.A., Mendes, A.M.S. (2022a). Disturbance observer based predictive current control of six-phase permanent magnet synchronous machines for the mitigation of steady-state errors and current harmonics. *IEEE Transactions on Industrial Electronics*, 69(1), 130–140.

Gonçalves, P.F.C., Cruz, S.M.A., Mendes, A.M.S. (2022b). Multistage predictive current control based on virtual vectors for the reduction of current harmonics in six-phase PMSMs. *IEEE Transactions on Energy Conversion*, 36(2), 1368–1377.

Gonçalves, P.F.C., Cruz, S.M.A., Mendes, A.M.S. (2022c). Online diagnostic method for the detection of high-resistance connections and open-phase faults in six-phase PMSM drives. *IEEE Transactions on Industry Applications*, 58(1), 345–355.

Gonzalez-Prieto, I., Duran, M.J., Aciego, J.J., Martin, C., Barrero, F. (2018). Model predictive control of six-phase induction motor drives using virtual voltage vectors. *IEEE Transactions on Industrial Electronics*, 65(1), 27–37.

Gritli, Y., Tani, A., Rossi, C., Casadei, D. (2019). Assessment of current and voltage signature analysis for the diagnosis of rotor magnet demagnetization in five-phase AC permanent magnet generator drives. *Mathematics and Computers in Simulation*, 158, 91–106.

Gritli, Y., Mengoni, M., Rizzoli, G., Rossi, C., Tani, A., Casadei, D. (2020). Rotor magnet demagnetisation diagnosis in asymmetrical six-phase surface-mounted AC PMSM drives. *IET Electric Power Applications*, 14(10), 1747–1755.

Guo, H., Guo, S., Xu, J., Tian, X. (2021). Power switch open-circuit fault diagnosis of six-phase fault tolerant permanent magnet synchronous motor system under normal and fault-tolerant operation conditions using the average current Park's vector approach. *IEEE Transactions on Power Electronics*, 36(3), 2641–2660.

Hu, R., Wang, J., Sen, B., Mills, A.R., Chong, E., Sun, Z. (2017). PWM ripple currents based turn fault detection for multiphase permanent magnet machines. *IEEE Transactions on Industry Applications*, 53(3), 2740–2751.

Immovilli, F., Bianchini, C., Lorenzani, E., Bellini, A., Fornasiero, E. (2015). Evaluation of combined reference frame transformation for interturn fault detection in permanent-magnet multiphase machines. *IEEE Transactions on Industrial Electronics*, 62(3), 1912–1920.

Karttunen, J., Kallio, S., Honkanen, J., Peltoniemi, P., Silventoinen, P. (2017). Partial current harmonic compensation in dual three-phase PMSMs considering the limited available voltage. *IEEE Transactions on Industrial Electronics*, 64(2), 1038–1048.

Levi, E. (2008). Multiphase electric machines for variable-speed applications. *IEEE Transactions on Industrial Electronics*, 55(5), 1893–1909.

Levi, E., Bojoi, R., Profumo, F., Toliyat, H., Williamson, S. (2007). Multiphase induction motor drives – A technology status review. *IET Electric Power Applications*, 1(4), 489–516.

Liu, Z., Fang, L., Jiang, D., Qu, R. (2022). A machine-learning-based fault diagnosis method with adaptive secondary sampling for multiphase drive systems. *IEEE Transactions on Power Electronics*, 37(8), 8767–8772.

Moghadam, F.K. and Nejad, A.R. (2020). Evaluation of PMSG-based drivetrain technologies for 10-MW floating offshore wind turbines: Pros and cons in a life cycle perspective. *Wind Energy*, 23(7), 1542–1563.

Munim, W.N.W.A., Duran, M.J., Che, H.S., Bermudez, M., Gonzalez-Prieto, I., Rahim, N.A. (2017). A unified analysis of the fault tolerance capability in six-phase induction motor drives. *IEEE Transactions on Power Electronics*, 32(10), 7824–7836.

Nakao, N. and Akatsu, K. (2014). Suppressing pulsating torques: Torque ripple control for synchronous motors. *IEEE Industry Applications Magazine*, 20(6), 33–44.

Ogidi, O.O., Khan, A., Dehnavifard, H. (2020). Deployment of onshore wind turbine generator topologies: Opportunities and challenges. *International Transactions on Electrical Energy Systems*, 30(5), e12308.

Peng, X., Liu, Z., Jiang, D. (2021). A review of multiphase energy conversion in wind power generation. *Renewable and Sustainable Energy Reviews*, 147, 111172.

Prieto-Araujo, E., Junyent-Ferré, A., Lavernia-Ferrer, D., Gomis-Bellmunt, O. (2015). Decentralized control of a nine-phase permanent magnet generator for offshore wind turbines. *IEEE Transactions on Energy Conversion*, 30(3), 1103–1112.

Ren, Y. and Zhu, Z.-Q. (2015). Enhancement of steady-state performance in direct-torque-controlled dual three-phase permanent-magnet synchronous machine drives with modified switching table. *IEEE Transactions on Industrial Electronics*, 62(6), 3338–3350.

Salehifar, M., Salehi Arashloo, R., Moreno-Eguilaz, M., Sala, V., Romeral, L. (2015). Observer-based open transistor fault diagnosis and fault-tolerant control of five-phase permanent magnet motor drive for application in electric vehicles. *IET Power Electronics*, 8(1), 76–87.

Salmasi, F.R. (2017). A self-healing induction motor drive with model free sensor tampering and sensor fault detection, isolation, and compensation. *IEEE Transactions on Industrial Electronics*, 64(8), 6105–6115.

Telsnig, T. (2021). Wind energy technology development report 2020. Report, EUR 30503 EN.

Trabelsi, M., Nguyen, N.K., Semail, E. (2016). Real-time switches fault diagnosis based on typical operating characteristics of five-phase permanent-magnetic synchronous machines. *IEEE Transactions on Industrial Electronics*, 63(8), 4683–4694.

Wang, B., Wang, J., Griffo, A., Sen, B. (2018). Stator turn fault detection by second harmonic in instantaneous power for a triple-redundant fault-tolerant PM drive. *IEEE Transactions on Industrial Electronics*, 65(9), 7279–7289.

Wang, X., Wang, Z., Xu, Z., Cheng, M., Wang, W., Hu, Y. (2019). Comprehensive diagnosis and tolerance strategies for electrical faults and sensor faults in dual three-phase PMSM drives. *IEEE Transactions on Power Electronics*, 34(7), 6669–6684.

Wang, X., Wang, D., Zhu, S., Liu, Y., Zhao, H. (2021). Research on maximum power tracking control method of 10MW MVSPMSG based on neutral point potential balance. *IEEE Transactions on Applied Superconductivity*, 31(8), 1–4.

Williams, R., Zhao, F., Lee, J. (2022). GWEC: Global offshore wind report 2022. Report, GWEC.

Xia, Y., Fletcher, J., Finney, S., Ahmed, K., Williams, B. (2011). Torque ripple analysis and reduction for wind energy conversion systems using uncontrolled rectifier and boost converter. *IET Renewable Power Generation*, 5(5), 377–386.

Yao, G., Pang, S., Ying, T., Benbouzid, M., Ait-Ahmed, M., Benkhoris, M.F. (2020). VPSO-SVM-based open-circuit faults diagnosis of five-phase marine current generator sets. *Energies*, 13(22), 6004.

Yaramasu, V. and Wu, B. (2016). *Model Predictive Control of Wind Energy Conversion Systems*, 1st edition. John Wiley & Sons, New York.

Yaramasu, V., Wu, B., Sen, P.C., Kouro, S., Narimani, M. (2015). High-power wind energy conversion systems: State-of-the-art and emerging technologies. *Proceedings of the IEEE*, 103(5), 740–788.

Yaramasu, V., Dekka, A., Durán, M.J., Kouro, S., Wu, B. (2017). PMSG-based wind energy conversion systems: Survey on power converters and controls. *IET Electric Power Applications*, 11(6), 956–968.

Yepes, A.G., Gonzalez-Prieto, I., Lopez, O., Duran, M.J., Doval-Gandoy, J. (2022a). A comprehensive survey on fault tolerance in multiphase AC drives. Part 2: Phase and switch open-circuit faults. *Machines*, 10(3), 221.

Yepes, A.G., Lopez, O., Gonzalez-Prieto, I., Duran, M.J., Doval-Gandoy, J. (2022b). A comprehensive survey on fault tolerance in multiphase AC drives. Part 1: General overview considering multiple fault types. *Machines*, 10(3), 208.

Zhao, Y. and Lipo, T.A. (1995). Space vector PWM control of dual three-phase induction machine using vector space decomposition. *IEEE Transactions on Industry Applications*, 31(5), 1100–1109.

Zheng, L., Fletcher, J.E., Williams, B.W., He, X. (2011). A novel direct torque control scheme for a sensorless five-phase induction motor drive. *IEEE Transactions on Industrial Electronics*, 58(2), 503–513.

Zhou, D., Blaabjerg, F., Franke, T., Tønnes, M., Lau, M. (2015). Comparison of wind power converter reliability with low-speed and medium-speed permanent-magnet synchronous generators. *IEEE Transactions on Industrial Electronics*, 62(10), 6575–6584.

Zhu, Z. and Hu, J. (2013). Electrical machines and power-electronic systems for high-power wind energy generation applications. Part II: Power electronics and control systems. *COMPEL – The International Journal for Computation and Mathematics in Electrical and Electronic Engineering*, 32(1), 34–71.

3

Gearbox Fault Monitoring Using Induction Machine Electrical Signals

3.1. Introduction

Gearboxes are a critical element in numerous systems. In wind turbines (WTs), gearbox failure causes 6 days of downtime (Salameh et al. 2018), which stops the production task. Vibration analysis is the conventional method for monitoring electro-mechanical faults in rotating machinery. Its major drawback remains the additional cost and the installation size of the acquisition hardware as well as the complexity of the vibration sensor installation (radial, horizontal, on the bearing housings, on the machine, etc.), which highly affects the quality of the obtained signal and consequently on the efficiency of the fault detection results, which increases the false alarms and the non-detection levels.

Recently, monitoring gear-based systems driven by an induction machine using electromechanical spectral signature is the subject of several research works. The motor current signature analysis (MCSA) technique solves the drawbacks of vibration analysis because the acquisition of the stator current is simpler and more robust. The stator current sensors can be at any location on the machine power line. In addition, Hall effect current sensors are sufficient to convert high stator currents of the machine to an exploitable signal by computers and processing units. In addition, the stator currents acquisition hardware is already installed for the control process (motor

---

Chapter written by Khmais BACHA and Walid TOUTI.

For a color version of all figures in this chapter, see www.iste.co.uk/benkhaderbouzid/fault.zip.

vector control, WT pitch control, etc.). The induction machine helps to observe the state of the associated elements such as the gear, pump, compressor, crusher, etc. In a WT drivetrain, the state of the gearbox teeth can be monitored by performing the spectral analysis of the generator stator current by comparing the energy of the faults frequency components to the healthy state (Touti et al. 2021).

In the first part of this chapter, we will present the gear fault effect on the stator current frequency behavior using the load torque and the magnetic flux in the air gap. An improved theoretical basis of the amplitude modulation (AM) and the frequency modulation (FM) effect of gear failure on stator current will be presented. In addition, the MCSA result based on Gotix experimental data is used for gear fault detection. In the second part, various transformation methods of the stator currents will be presented and tested using the same experimental data to show their advantages compared with the MCSA technique. An alternative method based on the discrete cosine transform (DCT) and the discrete sine transform (DST) will show its improvement for gear fault detection using the same stator current data.

3.2. Motor stator current signature approach

3.2.1. *Air gap magnetic flux density-based approach*

3.2.1.1. *Gear effect on the load torque*

Induction machine drivetrain can present various mechanical faults such as rotor eccentricity, bearings faults, shaft misalignment and driven gearbox faults. These faults induce load torque oscillation on the rotor shaft, which is expressed in the healthy state as follows (Blodt et al. 2006; Kia et al. 2009):

$$\begin{aligned} T_{lh}(t) &= T_0 + T_{osc}(t) \\ &= T_0 + T_i(t) + T_o(t) + T_m(t) \end{aligned} \qquad [3.1]$$

where T_0 is the average torque, $T_i(t)$, $T_o(t)$ and $T_m(t)$ are the input, output and meshing torque components expressed as follows:

$$T_i(t) = T_i \sin(\omega_i t + \varphi_i) \qquad [3.2]$$

$$T_o(t) = T_o \sin(\omega_o t + \varphi_o) \qquad [3.3]$$

$$T_m(t) = T_m \sin(\omega_m t + \varphi_m) \qquad [3.4]$$

where T_i, T_o and T_m are the amplitudes of the $T_i(t)$, $T_o(t)$ and $T_m(t)$.

Note that $\omega_i = 2\pi f_i$, $\omega_o = 2\pi f_o$ and $\omega_m = 2\pi f_m$, where f_i, f_o and f_m are the input, output and meshing frequency of the gearbox.

In case of a gearbox fault, the load torque expression becomes:

$$\begin{aligned} T_{lf}(t) &= T_{lh}(t) + T_f(t) \\ &= T_0 + T_i(t) + T_o(t) + T_m(t) + T_f(t) \end{aligned} \qquad [3.5]$$

where T_f is the fault-related torque which can be expressed by:

$$T_f(t) = T_f \sin(\omega_f t + \varphi_f) \qquad [3.6]$$

where $\omega_f = 2\pi f_f$ and f_f is the frequency of the defective gear element.

The mechanical equation of the induction machine rotor shaft is given by:

$$\sum T(t) = T_{motor}(t) - T_{load}(t) = J \frac{d\omega_R(t)}{dt} \qquad [3.7]$$

where $T_0 = T_{motor}(t)$ and ω_R is the rotor speed is obtained as follows:

$$\omega_R(t) = -\frac{1}{J} \int_{t_0}^{t} \left[T_i(\tau) + T_o(\tau) + T_m(\tau) + T_f(\tau) \right] d\tau + \omega_{R0} \qquad [3.8]$$

Then, the mechanical angle θ_r can be computed as:

$$\begin{aligned} \theta_R(t) &= \int_{t_0}^{t} \omega_R(\tau) d\tau \\ &= \frac{1}{p} \psi(t) + \omega_{R0} t + \theta_{R0} \end{aligned} \qquad [3.9]$$

where p is pole pairs.

As it is well known, in the case of induction motor without a gearbox

$$\theta_R(t) = \omega_{R0}t + \theta_{R0} \qquad [3.10]$$

The term $\psi(t)$ in the faulty case is given by:

$$\psi_f(t) = \beta_i \sin(\omega_i t + \varphi_i) + \beta_o \sin(\omega_o t + \varphi_o) \\ + \beta_m \sin(\omega_m t + \varphi_m) + \beta_f \sin(\omega_f t + \varphi_f) \qquad [3.11]$$

where: $\beta_i = \dfrac{pT_i}{J\omega_i^2}$ [3.12]

$$\beta_o = \dfrac{pT_o}{J\omega_o^2} \qquad [3.13]$$

$$\beta_m = \dfrac{T_m}{J\omega_m^2} \qquad [3.14]$$

$$\beta_f = \dfrac{T_f}{J\omega_f^2} \qquad [3.15]$$

In the case of the input shaft fault, $\omega_f = \omega_i$. As a result,

$$\psi_f(t) = 2\beta_i \sin(\omega_i t + \varphi_i) + \beta_o \sin(\omega_o t + \varphi_o) + \beta_m \sin(\omega_m t + \varphi_m) \qquad [3.16]$$

It has been shown that the fault case is characterized by the increase in the input shaft-related component, which can reflect the fault effect in the other signals.

3.2.1.2. Gear effect on the magnetomotive force

Torsional vibration on corresponding rotating shafts is presented by the mechanical angle θ_r variation, which affects the rotor magnetomotive force F_r expressed in the rotor reference frame as follows:

$$F_r^R(\theta^r, t) = F_r \cos(p\theta^r - s\omega_s t) \qquad [3.17]$$

The transformation between the rotor reference frame and the stator reference frame is given by:

$$\theta^s = \theta^r + \theta_r(t)$$
$$= \theta^r + \psi(t) + \frac{1-s}{p}\omega_s t + \theta_{r0} \qquad [3.18]$$

where $\omega_s = 2\pi f_s$, f_s is the supply frequency of the induction machine.

Then, the rotor magnetomotive force in the stator reference frame becomes:

$$F_r^s(\theta^s,t) = F_r \cos\left(p\theta^s - \omega_s t - \psi(t)\right) \qquad [3.19]$$

Load torque oscillation of the gear characteristic frequencies leads to phase modulation of the magnetomotive force created in the rotor by the term $\psi(t)$.

The stator magnetomotive force in the stator reference frame is given by:

$$F_s^s(\theta^s,t) = F_s \cos\left(p\theta^s - \omega_s t\right) \qquad [3.20]$$

The magnetic flux density in the air gap can be modified either by the effect of the variation of the air gap length and its permeance or by the effect of the load torque variation. To study the last effect, we consider that the machine has a constant air gap and that its permeance is constant. Then, the total magnetic flux density in the air gap is given by:

$$B(\theta^s,t) = \left[F_r^s(\theta^s,t) + F_s^s(\theta^s,t)\right]\lambda_0$$
$$= B_s \cos\left(p\theta^s - \omega_s t\right) + B_r \cos\left(p\theta^s - \omega_s t - \psi(t)\right) \qquad [3.21]$$

where λ_0 is the air gap permeance and assuming that the initial angle $\theta_{r0} = 0$. By integrating the flux density $B(\theta^s, t)$, the air gap flux $\phi(t)$ in each phase can be written as:

$$\phi(t) = \phi_s \cos(\omega_s t + \varphi_s) + \phi_r \cos(\omega_s t + \psi(t) + \varphi_r) \qquad [3.22]$$

3.2.1.3. *Gear effect on the electromotive force*

Before expressing the stator current of the linked machine to the faulty gear, the electromotive force (EMF) is an induced voltage E in the stator winding, which can be derived from the magnetic flux of [3.22]:

$$E(t) = -\phi_s \omega_s \sin(\omega_s t + \varphi_s) + \phi_r \cos'\left[\omega_s t + \psi(t) + \varphi_r\right]$$
$$= -\underbrace{\phi_s \omega_s \sin(\omega_s t + \varphi_s)}_{\text{AM effect}} - \underbrace{\phi_r \left[\omega_s + \psi'(t)\right] \sin(\omega_s t + \psi(t) + \varphi_r)}_{\text{FM effect}} \quad [3.23]$$

3.2.1.4. *Gear modulation effect on the motor stator current*

The fundamental differential equation of the induction machine in the stator is given by:

$$v(t) = R_s i(t) + \frac{d\phi(t)}{dt} \quad [3.24]$$

The air gap flux $\phi(t)$ produces components in the stator current of the induction machine. Considering that the stator voltage is maintained constant for each phase, the frequency content of the stator current $i(t)$ is then assumed to be the same as that of the EMF; it can be expressed as follows:

$$i_{saf}(t) = i_{sah}(t) + i_{ra\,mod}(t)$$
$$= \underbrace{I_{sah} \omega_s \cos(\omega_s t)}_{\text{AM effect}} + \underbrace{I_{ra\,mod}\left[\omega_s + \psi'(t)\right] \times \cos\left[\omega_s t + \psi(t)\right]}_{\text{FM effect}} \quad [3.25]$$

$$i_{saf}(t) = i_{sah}(t) + \underbrace{I_{sa\,mod} \omega_s \cos\left[\omega_s t + \psi(t)\right]}_{i_{raAM}(t)} + \underbrace{I_{sa\,mod} \psi'(t) \cos\left[\omega_s t + \psi(t)\right]}_{i_{raFM}(t)} \quad [3.26]$$

$$i_{saf}(t) = i_{sah}(t) + i_{raAM}(t) + i_{raFM}(t) \quad [3.27]$$

The geared motor stator current consists of the combination of two components, an unmodulated component $I_{sah}(t)$ resulting from the magnetomotive force created in the stator and the modulated component

$I_{ramod}(t)$ resulting from the magnetomotive force created in the rotor; in addition to the modulated terms, this modulation is due to the oscillation of the load torque caused by the vibrations of the gear shafts.

The component $i_{arAM}(t)$ was studied in Kia et al. (2009) to prove the gear fault modulation effect. In this chapter, the second component $i_{arFM}(t)$ is studied for the same subject. The term $i_{arFM}(t)$ was neglected in Blodt et al. (2006) and Kia et al. (2009), since the modulation index $\beta_i, \beta_o\ldots \ll 1$. In Feng et al. (2019), the authors affirm that the condition $\beta_i, \beta_o\ldots \ll 1$ is not usually the case. In the proposed model, the second term in the AM part is not neglected. As such, the model is generalized to accommodate a more general case. More importantly, it motivates both amplitude and frequency demodulation analyses for gear fault feature extraction. Replacing the expression of $\psi'(t)$ in $i_{arFM}(t)$, we obtain:

$$i_{arFM}(t) = I_{ar}\cos[\omega_s t + \psi(t)]\{2\omega_i\beta_i\cos(\omega_i t + \varphi_i') \\ + \omega_o\beta_o\cos(\omega_o t + \varphi_o') + \omega_m\beta_m\cos(\omega_m t + \varphi_m')\} \quad [3.28]$$

Neglecting the initial angles φ',

$$i_{arFM}(t) = \frac{I_{ar}}{2}\{2\omega_i\beta_i[\cos[(\omega_s + \omega_i)t + \psi(t)] + \cos[(\omega_s - \omega_i)t + \psi(t)]] \\ + \omega_o\beta_o[\cos[(\omega_s + \omega_o)t + \psi(t)] + \cos[(\omega_s - \omega_o)t + \psi(t)]] \\ + \omega_m\beta_m[\cos[(\omega_s + \omega_m)t + \psi(t)] + \cos[(\omega_s - \omega_m)t + \psi(t)]]\} \quad [3.29]$$

Using a trigonometrical approach and factorization, $i_{r2}(t)$ becomes:

$$i_{arFM}(t) = [2\sin(\omega_i t) + \sin(\omega_o t) + \sin(\omega_m t)]I_{ar}\cos(\omega_s t + \psi(t)) \quad [3.30]$$

Using the following Jacobi Anger expansion:

$$\exp(j\beta\cos\gamma) = \sum_{k=-\infty}^{+\infty} j^k J_k(\beta)\exp(jk\gamma) \quad [3.31]$$

We obtain:

$$i_{arFM}(t) = [2\sin(\omega_i t) + \sin(\omega_o t) + \sin(\omega_m t)]$$
$$\times I_{ar}\exp[j\omega_s t] \times \sum_{k=-\infty}^{+\infty} j^k J_k(2\beta_i)\exp[jk(\omega_i t + \varphi'_i)]$$
$$\times \sum_{k=-\infty}^{+\infty} j^k J_k(\beta_o)\exp[jk(\omega_o t + \varphi'_o)] \qquad [3.32]$$
$$\times \sum_{k=-\infty}^{+\infty} j^k J_k(\beta_m)\exp[jk(\omega_m t + \varphi'_m)]$$

The Fourier transform of $i_{arFM}(t)$ is given as

$$I_{arFM}(f) = [\delta(f - f_i) + \delta(f - f_o) + \delta(f - f_m)]$$
$$* I_{ar}\delta(f - f_s) * \sum_{k=-\infty}^{+\infty} J_k(2\beta_i)\delta(f - f_i)$$
$$* \sum_{k=-\infty}^{+\infty} J_k(\beta_o)\delta(f - f_o) * \sum_{k=-\infty}^{+\infty} J_k(\beta_m)\delta(f - f_m) \qquad [3.33]$$

where (*) is the convolution product operator and $\delta(f)$ is the Dirac function.

The convolution between the second term of the product and the following terms gives a modulation frequency components related to the input gear shaft frequency f_i, the output gear shaft f_o and the meshing frequency f_m, which can be expressed by:

$$|f_s \pm kf_i| \qquad [3.34]$$

$$|f_s \pm kf_o| \qquad [3.35]$$

$$|f_s \pm kf_m| \qquad [3.36]$$

where k is a real integer.

Gearbox Fault Monitoring Using Induction Machine Electrical Signals 97

The first term of the product in [3.33] leads to the addition of the amplitudes of the spectral lines of the rotation frequencies of the drivetrain elements; this coefficient amplifies the amplitudes of the modulation components in the previous equation.

The component $(f_s \pm f_i)$ reflects the fault information in the input shaft running with the frequency f_i. In case of the pinion fault, this component is amplified by 2. This result will be verified in section 3.4.

3.2.2. *Magnetizing current approach*

It was previously shown that the torsional vibrations generated by the gear elements produce both amplitude modulation and frequency modulation in the stator currents of the machine. Consider that the stator current in each phase is the sum of a magnetizing or reactive component $I_{sM}(t)$ almost equal to the no-load current and an active component producing torque $I_{sT}(t)$ (Figure 3.1).

These terms are sinusoidal. Both components also present the same frequency modulation effect created by the gear elements. Assuming the inference in Yacamini et al. (1998), the high vibration frequencies are damped by the machine structure.

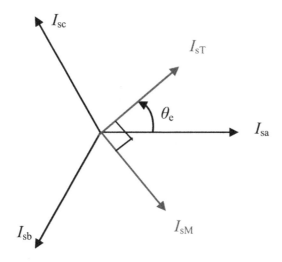

Figure 3.1. *Vector diagram of current components in the induction machine*

As a result, the relatively high meshing frequency will be neglected in the stator currents and consequently in the magnetizing and active components, which will be expressed by:

$$i_{sM}(t) = I_{sM0} + A_{sMi}\sin(2\pi f_i t + \varphi_{Mi}) + A_{sMo}\sin(2\pi f_o t + \varphi_{Mo})$$
$$i_{sT}(t) = I_{sT0} + A_{sTi}\sin(2\pi f_i t + \varphi_{Ti}) + A_{sTo}\sin(2\pi f_o t + \varphi_{To})$$

[3.37]

Stator currents can be expressed through the following transformation:

$$\begin{bmatrix} i_{sa}(t) \\ i_{sb}(t) \\ i_{sc}(t) \end{bmatrix} = \begin{bmatrix} \sin(2\pi f_s t) & \cos(2\pi f_s t) \\ \sin(2\pi f_s t - 2\pi/3) & \cos(2\pi f_s t - 2\pi/3) \\ \sin(2\pi f_s t + 2\pi/3) & \cos(2\pi f_s t + 2\pi/3) \end{bmatrix} \begin{bmatrix} i_{sM}(t) \\ i_{sT}(t) \end{bmatrix}$$

[3.38]

A stator current is then written as:

$$\begin{aligned} i_{arFM}(t) &= \left[i_{sM}(t) \times \sin(2\pi f_s t)\right] + \left[i_{sT}(t) \times \cos(2\pi f_s t)\right] \\ &= I_0 \sin(2\pi f_s t + \varphi_0) \\ &+ \left(\frac{A_{sMi} + A_{sTi}}{2}\right)\cos(2\pi(f_s - f_i)t - \varphi_{Mi}) \\ &+ \left(\frac{A_{sMi} - A_{sTi}}{2}\right)\cos(2\pi(f_s + f_i)t + \varphi_{Mi}) \\ &+ \left(\frac{A_{sMo} + A_{sTo}}{2}\right)\cos(2\pi(f_s - f_o)t - \varphi_{Mo}) \\ &+ \left(\frac{A_{sMo} - A_{sTo}}{2}\right)\cos(2\pi(f_s + f_o)t + \varphi_{Mo}) \end{aligned}$$

[3.39]

where $I_{sM0} = I_0 \cos\varphi_0$; $I_{sT0} = I_0 \sin\varphi_0$ and $\varphi_0 = \arctan(I_{sT0}/I_{sM0})$ assuming $\varphi_{Mi} = \varphi_{Ti}$.

This result justifies equations [3.34] and [3.35], where the stator current presents frequency modulations in the components $(f_s \pm f_i)$ and $(f_s \pm f_o)$ due to the gears torsional vibration.

3.3. Wound rotor current signature approach

This case is attracted to the WT application, where the doubly fed induction generator is driven by the high-speed shaft of the gearbox. The rotor current can be used for gear fault monitoring. The frequency effect of the gearbox on the rotor current can be shown using the following expression of the flux vector in the stator λ_s:

$$\lambda_s = L_s i_s + L_m i_r e^{j\theta_r} \qquad [3.40]$$

where L_s is the stator inductance and L_m is the mutual inductance.

The rotor current vector is given as follows:

$$i_r = \frac{\lambda_s - L_s i_s}{L_m} e^{-j\omega_r t} \qquad [3.41]$$

The representation of this equation in the stationary frame (α, β) developed by (Cheng et al. 2017) gives:

$$\begin{aligned} i_r &= \frac{\lambda_{s\alpha} + j\lambda_{s\beta}}{L_m}(\cos\omega_r t - j\sin\omega_r t) - \frac{L_s}{L_m}(i_{s\alpha} + ji_{s\beta})(\cos\omega_r t - j\sin\omega_r t) \\ &= \frac{\lambda_{s\alpha}\cos\omega_r t + \lambda_{s\beta}\sin\omega_r t - L_s(i_{s\alpha}\cos\omega_r t + i_{s\beta}\sin\omega_r t)}{L_m} \\ &\quad + j\frac{\lambda_{s\beta}\cos\omega_r t - \lambda_{s\alpha}\sin\omega_r t - L_s(i_{s\beta}\cos\omega_r t + i_{s\alpha}\sin\omega_r t)}{L_m} \\ &= i_{r\alpha} + ji_{r\beta} \end{aligned} \qquad [3.42]$$

where $\lambda_{s\alpha}$ and $\lambda_{s\beta}$ are the stator flux components along α and β axis, respectively, and $i_{s\alpha}$ and $i_{s\beta}$ are the stator current components along α and β axis respectively. Here, $\omega_r = 2\pi f_r$, where f_r is the rotor current fundamental frequency. Using the transformation from the two-phase reference frame to the three phases reference frame, the rotor current can be expressed as follows:

$$\begin{bmatrix} i_{ra} \\ i_{rb} \\ i_{rc} \end{bmatrix} = \begin{bmatrix} 1 & 0 \\ -\frac{1}{2} & \frac{\sqrt{3}}{2} \\ -\frac{1}{2} & -\frac{\sqrt{3}}{2} \end{bmatrix} \begin{bmatrix} i_{r\alpha} \\ i_{r\beta} \end{bmatrix}$$ [3.43]

The rotor current can be written as:

$$i_{ra}(t) = \frac{1}{L_m}\{\lambda_{s\alpha}\cos\omega_r t + \lambda_{s\beta}\sin\omega_r t - L_s(i_{s\alpha}\cos\omega_r t + i_{s\beta}\sin\omega_r t)\}$$

$$= \frac{1}{L_m}\left\{\left(\lambda_{sa} - \frac{1}{2}\lambda_{sb} - \frac{1}{2}\lambda_{sc}\right)\cos\omega_r t + \left(\frac{\sqrt{3}}{2}\lambda_{sb} - \frac{\sqrt{3}}{2}\lambda_{sc}\right)\sin\omega_r t\right.$$ [3.44]

$$\left. -L_s\left(\left(i_{sa} - \frac{1}{2}i_{sb} - \frac{1}{2}i_{sc}\right)\cos\omega_r t + \left(\frac{\sqrt{3}}{2}i_{sb} - \frac{\sqrt{3}}{2}i_{sc}\right)\sin\omega_r t\right)\right\}$$

Based on the fundamental relation [3.24], the stator flux in phase "a" is expressed by:

$$\lambda_{sa} = \int_t \left[v_{sa}(t) - R_s i_{sa}(t)\right]dt$$ [3.45]

From this equation, it is observed that the frequency content of the stator flux is the same as that of stator current $i_{sa}(t)$ and stator voltage $v_{sa}(t)$, which is $(f_s \pm f_i)$ and $(f_s \pm f_o)$.

The expression [3.44] of the rotor current in phase "a" is the product of $\sin(\omega_s \pm \omega_j)\cos(\omega_r t)$, $\sin(\omega_s \pm \omega_j)\sin(\omega_r t)$ or $\cos(\omega_s \pm \omega_j)\cos(\omega_r t)$ using the trigonometric equalities:

$$\frac{1}{2}\left[\cos(a-b) \pm \cos(a+b)\right] \text{ and } \frac{1}{2}\left[\sin(a-b) + \sin(a+b)\right]$$

And knowing that $a = \omega_s \pm \omega_j$ and $b = \omega_r$, the frequency components of the rotor current in phase "a" can be expressed by:

$$\begin{aligned} &|f_s \pm f_i| \\ &|f_s \pm f_o| \\ &\left|f_s \pm f_i \pm \frac{\omega_r}{2\pi}\right| \\ &\left|f_s \pm f_o \pm \frac{\omega_r}{2\pi}\right| \end{aligned} \qquad [3.46]$$

3.4. Experimental results

3.4.1. *MCSA for geared motor fault diagnosis*

In geared motor applications, the stator current presents frequency modulation components across the supply frequency according to equations [3.34]–[3.36]. These results have been confirmed by several works (Kar et al. 2006; Kia et al. 2009a, 2009b; Feki et al. 2013). In this section, the spectral analysis of the stator current data of the Gotix test bench (Figure 3.2) will be presented. This test rig includes a 55 kW Leroy Somer induction motor (Figure 3.2(a)) powered by a 400 V/50 Hz three-phase supply and a 54 kW Leroy Somer DC generator as load, through a single-stage, parallel shafts spur gearbox (Figure 3.2(b)).

Figure 3.2. *Gotix test rig: (a) motor side; (b) gearbox side*

The gearbox and the induction motor parameters are listed in Table 3.1.

All stator currents and vibration data are sampled synchronously using the Racal analyzer (Figure 3.3) with a 25 kHz of sampling frequency and 10 s of the time acquisition. The system is fully loaded at 170 Nm using a controlled load DC generator.

Parameter	Designation	Value	
Input gear (pinion) teeth	Z_i	57	
Output gear (wheel) teeth	Z_o	15	
Supply frequency (Hz)	f_s	50	
Induction machine pole pairs	P	4	
Input frequency (rotor frequency) (Hz)	$f_i = \dfrac{f_s}{p}$	Theoretical	Experimental (Figure 3.4)
		12.5	12.33
Output frequency (Hz)	f_o	47.5	46.83
Meshing frequency (Hz)	$f_m = Z_i f_i = Z_o f_o$	712.5	Missing

Table 3.1. Parameters and rotational frequencies of the geared motor

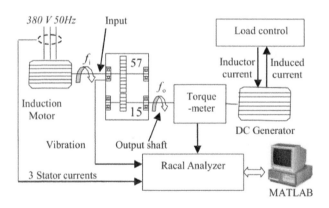

Figure 3.3. Gotix experimentation (Touti et al. 2016)

In the spectrum of Figure 3.4, some modulation frequency components appear across the fundamental frequency expressed by two terms. As a result, four significant components can be identified:

The term $|f_s \pm f_i|$ is detected in the healthy case (2807h) for $k = -1$ ($f_s - f_i = 37.7$ Hz) and for $k = +1$ ($f_s + f_i = 62.3$ Hz). Moreover, in the faulty case

(3465h), the levels of these components increase by 3.81 dB, which is predicted theoretically by the term $2\beta_i$. This result means that the input gear teeth are defective.

The term $|f_s \pm f_o|$ reflects the fault information on the output stage. However, in the healthy case (2807h) for $k = -1$, this component has a low amplitude ($f_s - f_o = 3.25$ Hz) and for $k = +1$ ($f_s + f_o = 96.6$ Hz). Likewise, in the faulty case (3465h), this component is absent.

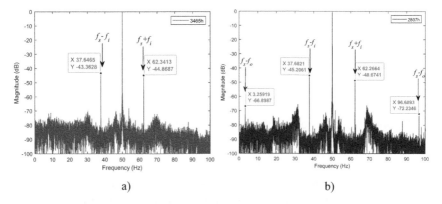

Figure 3.4. MCSA for the Gotix geared motor fault diagnosis: (a) healthy case (2807h); (b) fault case (3465h)

Modulation components	Values (Hz)	Magnitudes (dB)		Sensitivity (dB)
		Healthy case	Faulty case	
$f_s + f_i$	62.5	−48.67	−44.86	3.81
$f_s - f_i$	37.6	−45.2	−43.36	1.84
$f_s + f_o$	96.68	−72.23	Missing	Missing
$f_s - f_o$	3.25	−66.89	Missing	Missing

Table 3.2. Modulation components in the stator current for $k = 1$

3.4.2. MCSA for WT gearbox

The case of a 25 kW small WT was studied by Touti et al. (2021), a multistage gearbox in the WT drivetrain was investigated in two working conditions, with and without stripped pinion fault in the lay shaft. The squirrel-cage induction generator stator current was used as a faulty sensor. The current data were sampled at 4 kHz during an interval of 4 s. As shown in Figure 3.5, the spectral analysis of the generator stator current showed all shaft frequencies across the fundamental frequency f_e as a modulation effect (Table 3.3).

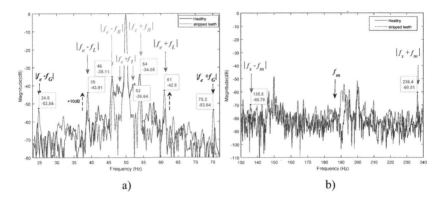

Figure 3.5. *Generator stator current spectrum of the WT (a) around the fundamental frequency and (b) around the meshing frequency (Touti 2022)*

It can be seen that the lay shaft-related component level $|f_e \pm f_L|$ is increased by 10 dB in the faulty case, knowing that f_L is the frequency of the defective pinion; this increase is due to the fault effect. Likewise, the modulation component expressed in [3.36] is detected for $k = \pm 1$ (Figure 3.5(b)) by an average increase of 8 dB.

3.4.3. *WT generator current processing*

Due to the multicomponent modulation and the background noise (Touti et al. 2021), gear fault frequencies can be hidden or weak in magnitude. In this subject, signal processing techniques can improve the frequency analysis of the generator's current signal.

The same data used from the previous WT are processed by the wavelet packets transform (WPT) and local mean decomposition (LMD). WPT decomposes the signal to several nodes into different frequency ranges. The node (4,1) corresponds to the range (125–250) Hz, which includes the meshing frequency f_{m2} = 186 Hz of the gearbox input shaft. By performing the FFT to this node (Figure 3.6(a)), the component is detected by an increase of 25 dB compared to the healthy case.

Related component	Practical (Hz)	Sideband (Hz)	Magnitude (dB) (healthy)	Magnitude (dB) (fault)	Fault Sensitivity (dB)
Turbine $\|f_e \pm f_T\|$	$\|50 \pm 2\|$	52	−39.64	−55.9	−16.26
		48	−40	−59.46	−19.46
Blade $\|f_e \pm f_B\|$	$\|50 \pm 4\|$	54	−34.05	−47.29	−13.24
		46	−35.11	−43.47	−8.36
Lay shaft $\|f_e \pm f_L\|$	$\|50 \pm 11\|$	61	−52.85	−42.8	+10.05
		39	−53.42	−43.91	+9.51
GMF $\|f_e \pm f_{m2}\|$	$\|50 \pm 186\|$	236	−69.29	−60.51	+8.78
		136	−76.87	−69.78	+7.09
Rotor $\|f_e \pm f_G\|$	$\|50 \pm 25.2\|$	75.2	−63.51	−53.64	+9.87
		24.8	−60.53	−52.54	+7.99

Table 3.3. *Experimental stator current modulation components of the WT gearbox (Touti 2022)*

The LMD is a demodulation technique that separates the mixed AM and the FM components from the original signal. The spectrum of the second amplitude modulated component AM$_2$ (Figure 3.6(b)) shows the frequency f_m and the related modulation components $|f_m \pm kf_L|$. Besides the processing methods, the transformations of the stator currents are an alternative to improve the spectral analysis.

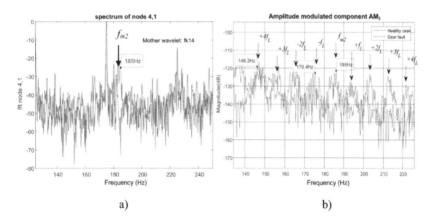

Figure 3.6. *Spectra of processed stator current of the WT: (a) spectrum of the node (4,1) using WPT; (b) spectrum of the second demodulated component (Touti et al. 2022)*

3.4.4. *Current transformations for geared motor fault diagnosis*

Except for the conventional technique (MCSA), various extended techniques such as the extended park vector approach (EPVA) and the symmetrical components method are used to improve the fault detection task. In this subject, EPVA is used recently to monitor numerous cases of induction motor mechanical faults (bearing faults (Zarei et al. 2009), broken bars faults (Bacha et al. 2012; Vaimann et al. 2012; Bouslimani et al. 2014), electricals faults (winding faults; Pires et al. 2009), WT faults (pitch system fault; Kandukuri et al. 2018) and blade imbalance faults (Yang et al. 2019)). The homopolar component is efficient in the case of unbalanced current fault detection.

In the next sections, each method is discussed separately.

3.4.4.1. *Extended Park vector approach*

3.4.4.1.1. The Park transform

The Park transform is used to model a three-phase system using a two-phase model for the order reduction of the machine equations, in particular for vector control. This is a change in the reference frame. The first two axes in the new basis are traditionally named *d*, *q*. The quantities transformed are generally currents, voltages or fluxes.

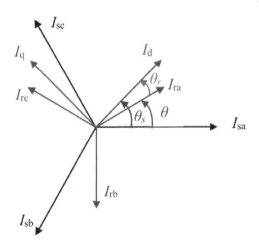

Figure 3.7. *Induction machine space vector diagram*

The transformation is given by:

$$\begin{bmatrix} i_d \\ i_q \\ i_0 \end{bmatrix} = P(\theta) \begin{bmatrix} i_{sa} \\ i_{sb} \\ i_{sc} \end{bmatrix} \qquad [3.47]$$

where i_{sa}, i_{sb} and i_{sc} are the stator currents in the stator reference frame,

$$\begin{cases} i_{sa}(t) = I_{sa} \cos(2\pi f_s t - \alpha_0) \\ i_{sb}(t) = I_{sb} \cos(2\pi f_s t - \dfrac{2\pi}{3}\alpha_0) \\ i_{sc}(t) = I_{sc} \cos(2\pi f_s t + \dfrac{2\pi}{3}\alpha_0) \end{cases} \qquad [3.48]$$

where i_f is the maximum amplitude of the fundamental supply line current and α_0 is the initial angle. And i_d, i_q and i_0 are the direct, quadrature and homopolar components. The passage matrix is given by:

$$P(\theta) = \sqrt{\dfrac{2}{3}} \begin{pmatrix} \cos(\theta) & \cos\left(\theta - \dfrac{2\pi}{3}\right) & \cos\left(\theta - \dfrac{4\pi}{3}\right) \\ -\sin(\theta) & -\sin\left(\theta - \dfrac{2\pi}{3}\right) & -\sin\left(\theta - \dfrac{4\pi}{3}\right) \\ \dfrac{1}{2} & \dfrac{1}{2} & \dfrac{1}{2} \end{pmatrix} \qquad [3.49]$$

The direct and quadrature currents (with $\theta = 0$) are expressed as follows:

$$\begin{cases} i_d(t) = \sqrt{\dfrac{2}{3}} i_a(t) - \sqrt{\dfrac{1}{6}} i_b(t) - \sqrt{\dfrac{1}{6}} i_c(t) \\ i_q(t) = \dfrac{1}{\sqrt{2}} i_b(t) - \dfrac{1}{\sqrt{2}} i_c(t) \end{cases} \qquad [3.50]$$

3.4.4.1.2. The extended Park approach

The EPVA uses the alternative component of Park's vector modulus, which is a complex vector that is collected by a real part $i_d(t)$ and an imaginary part $i_q(t)$. It is expressed as follows:

$$Pk = |i_d + ji_q| = \sqrt{i_d^2 + i_q^2} \qquad [3.51]$$

Zarei et al. (2009) and Silva et al. (2005) developed the expression of the stator current of the induction machine in the case of a bearing fault using the fault characteristic frequencies. This expression has been extended by Suo et al. (2018) and Kandukuri et al. (2018) for the case of a gear fault as expressed in [3.52]:

$$i_{sa}(t) = I_{sa}\cos(2\pi f_s t - \alpha_0)$$
$$+ i_{dl}\cos[2\pi(f_s - f_x)t - \beta_l] + i_{dr}\cos[2\pi(f_s + f_x)t - \beta_r]$$

$$i_{sb}(t) = I_{sb}\cos(2\pi f_s t - \alpha_0 - \frac{2\pi}{3})$$
$$+ i_{dl}\cos[2\pi(f_s - f_x)t - \beta_l - \frac{2\pi}{3}] + i_{dr}\cos[2\pi(f_s + f_x)t - \beta_r - \frac{2\pi}{3}] \quad [3.52]$$

$$i_{sc}(t) = I_{sc}\cos(2\pi f_s t - \alpha_0 + \frac{2\pi}{3})$$
$$+ i_{dl}\cos[2\pi(f_s - f_x)t - \beta_l + \frac{2\pi}{3}] + i_{dr}\cos[2\pi(f_s + f_x)t - \beta_r + \frac{2\pi}{3}]$$

where i_{dl} and i_{dr} are the maximum amplitudes of the current components at the frequencies (f_e-f_x) and (f_e+f_x), respectively; β_l and β_r are the initial phase angles and f_x is the frequency of torsional vibration induced by the rotation of the faulty gear element. In the case of a pinion fault, $f_x = f_i$, and in the case of a wheel fault, $f_x = f_o$. In the case of a gear fault, by replacing the expressions of the stator currents [3.52] in [3.50], the modulus of the Park's vector will be expressed by:

$$|Pk| = |i_d + ji_q| = \frac{2}{3}(i_0^2 + i_{dl}^2 + i_{dr}^2) + 3i_0 i_{dl}\cos(2\pi f_x t - \alpha + \beta_l)$$
$$+ 3i_0 i_{dr}^*\cos(2\pi f_x t + \alpha - \beta_r) + 3i_{dl} i_{dr}\cos(4\pi f_x t + \beta_l + \beta_r) \qquad [3.53]$$

The amplitudes of the gear fault components are multiplied by $3i_0$. Thus, these components are amplified, and consequently, their detection in the spectrum $Pk(f)$ will be improved. The flowchart of the EPVA is presented in Figure 3.8.

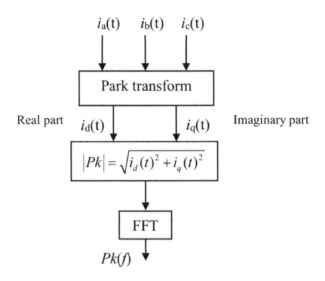

Figure 3.8. *Flowchart of the EPVA*

In Silva et al. (2005) and Zarei et al. (2009), the authors used the EPVA combined with the Hilbert transform to detect the rotor eccentricity fault (Salem et al. 2012) and the broken bars fault (Bacha et al. 2012). The gears fault profile was studied by Kia et al. (2015) using the instantaneous frequency of Park's vector. In this context, the authors show the mechanical impact generated by the gear tooth fault profile, which can be seen as fault-related frequencies in Park's vector instantaneous frequency. The study was verified on a 250 W squirrel-cage induction motor associated with a single-stage gearbox.

In Touti et al. (2018), authors investigated the frequency behavior of Park's vector using the experimental data from the Gotix test bench, which comprises a single-stage gearbox driven by a 55 kW induction motor. The spectral result is shown in Figure 3.9, where the input gearbox frequency f_i is amplified by the EPVA by 20.6 dB (−66.4 dB using MCSA and −45.75 dB using EPVA). The output frequency f_o detection is also enhanced.

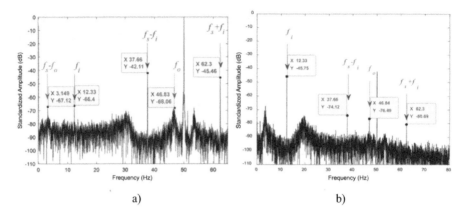

Figure 3.9. *Spectral analysis of the Gotix geared motor electrical signals: (a) MCSA result; (b) EPVA result*

The main drawback of the presented technique is the need for three sensors for the simultaneous acquisition of the three stator currents. However, the three sensors are required in case of imbalanced current fault detection. In this case, the Fortescue transform is called to define the zero sequence current expressed as follows:

$$i_0(t) = \frac{1}{\sqrt{3}}\left[i_{sa}(t) + i_{sb}(t) + i_{sc}(t)\right] \qquad [3.54]$$

This particular component is useful because the machine is star connected. Indeed, the current in the neutral is the image of the zero sequence component, and its acquisition requires a single sensor, which reduces the cost and size of the installation. In Amirat et al. (2013), this fault signature proved its effectiveness for the detection of WT generator bearing faults. In Touti et al. (2018), the fast Fourier transform (FFT) was performed on the zero sequence component of the stator currents of the Gotix bench. The spectra do not show a specific improvement in comparison with the MCSA technique.

3.4.4.2. *A fault signature based on the discrete cosine/sinus transform*

3.4.4.2.1. A complex-valued signal construction

The discrete cosine/sinus transform is primarily a 2D processing technique, often used for filtering images (denoising). Touti et al. (2018)

developed a gear fault diagnosis technique based on the dual cosine/sinus transform. A complex-valued signal (CVS) is obtained by a real component equal to the DCT of the input signal, while the imaginary part is the DST of the input signal. Similarly to the EPVA technique, the FFT is performed to the magnitude of the CVS to give the spectral content of the proposed technique. Moreover, the use of the DCT and the DST helps to attenuate the parasitic frequency components due to the boundary effect of the Fourier transform (theorem of Gibbs). In addition, compared to the discrete Fourier transform (DFT), this projection kernel is characterized by a strong energy compaction property allowing the information to be essentially carried by the low-frequency coefficients while allowing small high-frequency components to be discarded.

The proposed algorithm is developed as shown in Figure 3.10. Considering that the stator current signal $i_a(t)$ is sampled in N samples:

$$i = [i_n] = \begin{bmatrix} i_0 \\ i_1 \\ \cdot \\ \cdot \\ i_{N-1} \end{bmatrix} \quad [3.55]$$

The transformation matrices of the DCT and the DST are given, respectively, by C and S:

$$C = \begin{bmatrix} c_0^0 & c_0^1 & c_0^2 & \cdots & c_0^{N-1} \\ c_1^0 & c_1^1 & c_1^2 & & \\ c_2^0 & & & & \\ \cdot & & c_k^n & & \\ \cdot & & & & \\ c_{N-1}^0 & & & & c_{N-1}^{N-1} \end{bmatrix} \quad [3.56]$$

and

$$S = \begin{bmatrix} s_0^0 & s_0^1 & s_0^2 & \cdots & s_0^{N-1} \\ s_1^0 & s_1^1 & s_1^2 & & \\ s_2^0 & & & & \\ \cdot & & s_k^n & & \\ \cdot & & & & \\ s_{N-1}^0 & & & & s_{N-1}^{N-1} \end{bmatrix} \qquad [3.57]$$

where the coefficients of C and S are c_k^n and s_k^n expressed by:

$$c_k^n = \cos\left(\frac{(2n+1)k\pi}{2N}\right) \qquad [3.58]$$

and

$$s_k^n = \sin\left(\frac{(n+1)(k+1)\pi}{N+1}\right) \qquad [3.59]$$

The real coefficients of the DCT are computed as follows:

$$[I_k] = \sqrt{\frac{2w}{N}} \sum_{n=0}^{N-1} c_k^n . i_n = \sqrt{\frac{2w}{N}} \begin{bmatrix} c_0^0 & c_0^1 & c_0^2 & \cdots & c_0^{N-1} \end{bmatrix} \begin{bmatrix} i_0 \\ i_1 \\ \cdot \\ \cdot \\ i_{N-1} \end{bmatrix} \qquad [3.60]$$

where $w = \begin{cases} \dfrac{1}{2} & \text{if } k = 0 \\ 1 & \text{if } k \neq 0 \end{cases}$

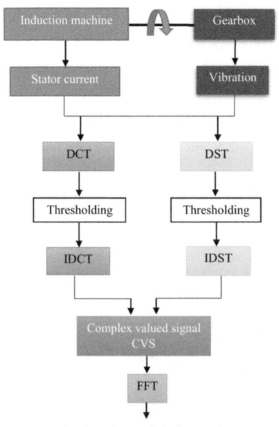

Figure 3.10. *Flowchart of the complex-valued signal CVS technique*

Similarly, the imaginary coefficients of the DST are computed as follows:

$$\underline{I} = \left[\underline{I_k}\right] = \begin{bmatrix} \underline{I_0} \\ \underline{I_1} \\ \underline{I_k} \\ . \\ . \\ I_{N-1} \end{bmatrix} = \sum_{k=0}^{N-1}\sum_{n=0}^{N-1} s_k^n . i_n \qquad [3.61]$$

It can be seen that the first term of the DCT ($k = 0$) corresponds to the mean of the input signal:

$$i_0 = \frac{(i_0 + i_1 + ... i_{N-1})}{\sqrt{N}} \quad [3.62]$$

After computing the DCT and the DST coefficients, the inverse transformation is performed using the inverse DCT and the inverse DST given by:

$$i_n = \sqrt{\frac{2w}{N}} \sum_{k=0}^{N-1} c_k^n I_k \quad [3.63]$$

and

$$i_n = \frac{2}{N+1} \sum_{k=0}^{N-1} s_k^n . I_k \quad [3.64]$$

The resulting CVS is the sequence y_n expressed by:

$$y = [y_n] = [y_0, y_1, ... y_{N-1}] \quad [3.65]$$

y_n is a complex vector of components:

$$\begin{cases} \text{Re}(y_n) = \frac{2}{N+1} \frac{2w}{N} \sum_{k=0}^{N-1} \sum_{n=0}^{N-1} \left(c_k^n . s_k^n . i_n \right) \\ \text{Im}(y_n) = \sqrt{\frac{2w}{N}} \sum_{k=0}^{N-1} \sum_{n=0}^{N-1} \left(c_k^n . s_k^n . i_n \right) \end{cases} \quad [3.66]$$

by putting the term G_N^w:

$$G_N^w = \sqrt{\frac{2w\left[4 + (N+1)^2\right]}{N(N+1)}} \quad [3.67]$$

The module of the vector $CVS(t)$ becomes:

$$|CVS(n)| = |y_n| = G_N^w \sum_{k=0}^{N-1} \sum_{n=0}^{N-1} \left(c_k^n . s_k^n . i_n \right) \quad [3.68]$$

The FFT of CVS is given by:

$$\overline{|Y_k|} = \left| \sum_{n=0}^{N-1} \sum_{k=0}^{N-1} \sum_{n=0}^{N-1} \left(G_N^w . c_k^n . s_k^n . i_n \right) e^{\frac{-2jkn\pi}{N}} \right| \quad [3.69]$$

3.4.4.2.2. Electrical signature enhancement

The precedent algorithm was applied to the stator current data of the Gotix geared motor. The level in the frequency domain increased compared to the MCSA result:

– the level of the input frequency f_i is increased by (+34.6 dB), −66.4 dB using the MCSA (Figure 3.11(a)) and −31.47 dB using the CVS (Figure 3.11(b));

– the level of the output frequency f_o is increased by +6 dB, −68.02 dB using the MCSA (Figure 3.12 (a)) and −61.89dB using the CVS (Figure 3.12(b)).

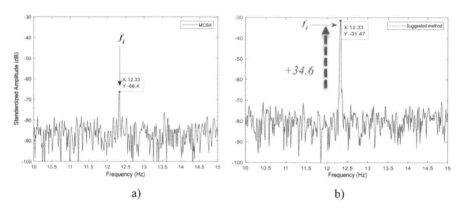

Figure 3.11. *Input frequency fi detection: (a) MCSA technique; (b) CVS spectrum*

This technique showed a good sensitivity for the gear characteristic frequencies detection as shown in Table 3.4.

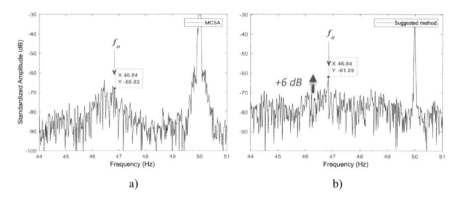

Figure 3.12. *Output frequency detection f_o: (a) MCSA technique; (b) CVS spectrum*

Frequency	Value (Hz)	MSCA magnitude (dB)	CVS magnitude (dB)	Sensibility (dB)	Figure
f_i	12.33	−66.4	−31.8	+34.6	3.11
f_o	46.84	−68.02	−61.89	+6	3.12

Table 3.4. *Comparison of the sensitivities of the MCSA technique versus the CVS technique for the gear characteristic frequencies*

3.5. Conclusion

As the gearbox is a critical component in various industrial applications, its reliability and condition monitoring is a basic tasks. This chapter presents a theoretical and experimental study of the gearbox fault's effect on the electrical signals of the driven induction machine. The induced gear fault frequencies are used as fault keys; in this way, the spectral analysis is established for each electrical signal such as the motor stator current, the stator current space vector and the discrete cosine sine-based transform signal. The last signal has shown its superiority by providing significant gains in the magnitudes of the gear characteristic frequencies.

3.6. Acknowledgments

The authors would like to thank Dr. Nadine Martin from the GIPSA-lab for providing data measurements and photos from the GOTIX test-bench designed for characterization of defects on electrically driven mechanical systems.

3.7. References

Amirat, Y., Choqueuse, V., Benbouzid, M. (2013). EEMD-based wind turbine bearing failure detection using the generator stator current homopolar component. *Mechanical Systems and Signal Processing*, 41(1), 667–678.

Bacha, K., Salem, S.B., Chaari, A. (2012). An improved combination of Hilbert and Park transforms for fault detection and identification in three-phase induction motors. *International Journal of Electrical Power & Energy Systems*, 43(1), 1006–1016.

Ben Salem, S., Bacha, K., Chaari, A. (2012). Support vector machine based decision for mechanical fault condition monitoring in induction motor using an advanced Hilbert-Park transform. *ISA Transactions*, 51(5), 566–572.

Blodt, M., Chabert, M., Regnier, J., Faucher, J. (2006). Mechanical load fault detection in induction motors by stator current time-frequency analysis. *IEEE Trans. Ind. Appl.*, 42(6), 1454–1463.

Bouslimani, S., Drid, S., Chrifi-Alaoui, L., Bussy, P., Ouriagli, M., Delahoche, L. (2014). An extended Park's vector approach to detect broken bars faults in Induction Motor. In *2014 15th International Conference on Sciences and Techniques of Automatic Control and Computer Engineering (STA)*. IEEE, Hammamet.

Cheng, F., Peng, Y., Qu, L., Qiao, W. (2017). Current-based fault detection and identification for wind turbine drivetrain gearboxes. *IEEE Transactions on Industry Applications*, 53(2), 878–887.

Feki, N., Clerc, G., Velex, P. (2013). Gear and motor fault modeling and detection based on motor current analysis. *Electric Power Systems Research*, 95, 28–37.

Feng, Z., Chen, X., Zuo, M.J. (2019). Induction motor stator current AM-FM model and demodulation analysis for planetary gearbox fault diagnosis. *IEEE Transactions on Industrial Informatics*, 15(4), 2386–2394.

Kandukuri, S.T., Van Khang, H., Robbsersmyr, K.G. (2018a). Multi-component fault detection in wind turbine pitch systems using extended Park's vector and deep autoencoder feature learning. In *2018 21st International Conference on Electrical Machines and Systems (ICEMS)*. Jeju.

Kar, C. and Mohanty, A.R. (2006). Monitoring gear vibrations through motor current signature analysis and wavelet transform. *Mechanical Systems and Signal Processing*, 20(1), 158–187.

Kia, S.H., Henao, H., Capolino, G.-A. (2009a). Analytical and experimental study of gearbox mechanical effect on the induction machine stator current signature. *IEEE Transactions on Industry Applications*, 45(4), 1405–1415.

Kia, S.H., Henao, H., Capolino, G.-A. (2009b). Torsional vibration effects on induction machine current and torque signatures in gearbox-based electromechanical system. *IEEE Transactions on Industrial Electronics*, 56(11), 4689–4699.

Pires, V.F., Amaral, T.G., Martins, J.F. (2009). Stator winding fault diagnosis in induction motors using the DQ current trajectory mass center. In *2009 35th Annual Conference of IEEE Industrial Electronics*. IEEE, Porto.

Salameh, J.P., Cauet, S., Etien, E., Sakout, A., Rambault, L. (2018). Gearbox condition monitoring in wind turbines: A review. *Mechanical Systems and Signal Processing*, 111, 251–264.

Silva, J.L.H. and Cardoso, A.J.M. (2005). Bearing failures diagnosis in three-phase induction motors by extended Park's vector approach. In *31st Annual Conference of IEEE Industrial Electronics Society. IECON 2005*. IEEE, Raleigh.

Suo, L., Liu, F., Xu, G., Wang, Z., Yan, W., Luo, A. (2018). Improved Park's vector method and its application in planetary gearbox fault diagnosis. In *2018 IEEE International Conference on Prognostics and Health Management (ICPHM)*. IEEE, Seattle.

Touti, W., Salah, M., Bacha, K., Amirat, Y., Chaari, A., Benbouzid, M. (2018). An improved electromechanical spectral signature for monitoring gear-based systems driven by an induction machine. *Applied Acoustics*, 141, 198–207.

Touti, W., Salah, M., Bacha, K., Chaari, A. (2022). Condition monitoring of a wind turbine drivetrain based on generator stator current processing. *ISA Transactions*, 128, 650–664.

Vaimann, T., Belahcen, A., Martinez, J., Kilk, A. (2012). Detection of broken bars in frequency converter-fed induction motor using Park's vector approach. In *2012 Electric Power Quality and Supply Reliability*. IEEE, Tartu.

Yacamini, R., Smith, K.S., Ran, L. (1998). Monitoring torsional vibrations of electro-mechanical systems using stator currents. *Journal of Vibration and Acoustics*, 120(1), 72–79.

Yang, D., Han, X., Fu, X., Du, J., Chen, G., Gao, Y., Wang, J. (2019). Blade imbalance fault diagnosis of DFIG based on current Park's transformation. In *2019 Prognostics and System Health Management Conference (PHM-Qingdao)*. IEEE, Qingdao.

Zarei, J. and Poshtan, J. (2009). An advanced Park's vectors approach for bearing fault detection. *Tribology International*, 42(2), 213–219.

4

Control of a Wind Distributed Generator for Auxiliary Services Under Grid Faults

4.1. Introduction

Given the significant growth in electricity consumption and population, the electrical network is becoming increasingly complex to deliver electricity from producers to consumers (Krim et al. 2022). This network includes power plants that generate electrical energy, high-voltage transmission lines that transport energy from distant sources to consumption centers and distribution lines that connect individual customers. However, this structural complexity negatively impacts the safety and quality of the electrical energy provided to consumers (Almaksour et al. 2021). The quality of electrical production is measured by its waveform, amplitude and frequency, and it is necessary to maintain these characteristics within established standards to ensure the quality of the production. Electrical energy quality can be compromised by internal incidents related to electrical receivers or external incidents related to physical phenomena, negatively affecting the three aforementioned characteristics. To address these electrical power quality issues, traditional adjustment and compensation methods are installed in each node of the distribution network. However, these conventional methods are generally slow and unable to maintain the stability of the distribution network (Krim et al. 2018a).

Chapter written by Youssef KRAIEM and Dhaker ABBES.

For a color version of all figures in this chapter, see www.iste.co.uk/benkhaderbouzid/fault.zip.

The liberalization of the electricity production market, coupled with growing public concern about global warming and climate change, has led to policies aimed at reducing the consumption of fossil fuels and greenhouse gas emissions by promoting the use of renewable energy sources (Krim et al. 2022). The integration of smart technologies and renewable production units into electrical energy transmission and distribution networks improves system flexibility and reduces the duration and cost of outages. Additionally, it increases utility revenue through faster power restoration and improved network capacity utilization. As renewable units are integrated closer to consumers, new energy flows are created, and new actors such as prosumers, electric vehicles and storage systems emerge (Cheikh-Mohamad et al. 2021). This leads to the development of energy communities, which bring together various actors within a distribution network. To manage energy exchanges between these actors, methods for controlling and managing energy are needed, leading to a convergence toward a smart grid.

A smart grid is an electrical network that comprises various operational and energy components, such as smart meters, smart appliances, sustainable energy generators and energy-efficient resources (Lamnatou et al. 2022). Effective control and regulation of electricity supply and distribution are crucial aspects of the smart grid. Decentralized renewable generators and smart grids have a significant role to play in the production and consumption of electrical energy. However, the intermittent nature and the prediction difficulty of renewable energy sources, particularly wind, complicate their use. Integrating these fluctuating sources into a perfectly controlled electrical network leads to power quality issues. Sudden changes in renewable energy generation disrupt power grid planning and the balance between supply and demand, making wind energy unreliable, even for isolated sites or direct consumption. Therefore, due to the variability and uncertainty of decentralized renewable generators, new challenges arise in power grid planning and operation. To mitigate these challenges and make the grid more accessible, flexible, economical and reliable, a monitoring and control methodology is proposed in this chapter. This methodology requires an intelligent system to ensure optimal energy flows for various purposes. Based on previous results, our approach focuses on exploring electrical microgrids (MGs) and their optimal and intelligent configurations. The concept of MG has emerged as a recent trend in innovation and research that highlights the contribution of renewable sources to system services (Krim et al. 2019).

A MG can be defined as a grid that is based on renewable power generation (RPG), and includes loads, generators and energy storage devices, such as batteries and supercapacitors (SCs), on low-voltage distribution systems (Krim et al. 2020). Energy storage units can address the issue of intermittence in wind production and meet the demands of loads. From a technical perspective, energy storage systems can provide a range of services to electrical systems. American and European studies have identified up to 30 services. For instance, supplying active power reserves to participate in grid frequency setting and adjustment mechanisms (Krim et al. 2020), adjusting grid voltage (via an inverter or modulation of active power injection on a distribution network), smoothing the active power injected into the grid/loads from RPG, which is inherently intermittent (Krim et al. 2021), managing occasional congestion on the grid (Krim et al. 2018b) and arbitrating on a market (value made up of forecasted differentials between low and high prices of daily markets) (Kofinas et al. 2018). In this context, different types of batteries such as nickel-metal hydride and lithium are being developed with the goal of having a lifespan of 20 years by 2030. Batteries are characterized by their specific energy, which is the amount of energy that can be delivered per unit mass or volume, and their specific capacity, which is the electrical charge that can be provided per unit mass. The lifespan of a battery is determined by its cyclability, or the average number of charge/discharge cycles it can perform. Although batteries have a high energy density, they have a slow dynamic response and low power density (Mendis et al. 2014). To improve the lifespan of batteries and reduce charge/discharge, SCs can be used. SCs are efficient storage systems that are used to smooth out rapid fluctuations in decentralized renewable energy production (Krim et al. 2018b). Hybridization between batteries and SCs is a promising solution to extend the life of batteries in renewable energy applications (Cabrane et al. 2017). This chapter will explore the combination of these two storage technologies.

The main issue with the storage system is overcharging and over-discharging. To address this, a power management algorithm (PMA) is necessary to manage the power flow between the various sources in the system. The PMA protects the battery/SC hybrid storage system (HSS) by maintaining the state of charge (SOC) within an acceptable range [SOC_{min}, SOC_{max}], thereby extending the battery's lifespan by reducing the number of charge/discharge cycles. PMAs have been widely studied in the literature, with rule-based PMS being the most commonly used (Krim et al. 2018c). However, in the context of renewable energy, rule-based PMA becomes complex due to multiple variables, and its all-or-nothing operation does not consider the physical aspect of the system (Yan et al. 2020). In such cases, fuzzy logic is a promising solution as it allows for more flexibility and human reasoning in the formulation of the problem, while avoiding mathematical rigidity and complexity. Therefore, this chapter proposes the use of fuzzy logic PMA.

The MG in this study consists of a wind generator, HSS, loads and a connection to the power grid. The MG can either operate in connected or standalone mode, depending on the state of the power grid. Energy storage devices help mitigate the issue of wind turbine intermittency by balancing the energy produced and consumed. Additionally, they support the stability of the electrical network by providing ancillary services such as voltage and frequency regulation. To achieve this, a control system for the connection to the power grid is necessary. The study by Taghizadeh et al. (2015) proposes a control strategy for frequency regulation in an islanded MG, while Rekik et al. (2013) suggest a droop control scheme to improve voltage and frequency levels using a wind generator and SC storage system. A generalized droop control method has been discussed in Krim et al. (2019), Shafiee et al. (2014) propose a robust networked control approach for secondary frequency and voltage control in islanded MGs, and Krim et al. (2018d) present a fuzzy logic droop control strategy to regulate power exchange with the grid and improve grid voltage levels.

This chapter focuses on the development of a fuzzy logic-based power supervisor and droop control for an electrical MG, taking into consideration loads and power grid uncertainties. The main contributions of this study are outlined as follows:

– The proposed model of the studied MG includes wind sources, a battery/SC hybrid stationary storage system, three-phase loads and connection to the power grid. The power grid is noted for its fluctuating frequency and voltage magnitude. The HSS operates within its power and capacity constraints to protect against maximum and minimum charging limits.

– A fuzzy logic-based power flow control approach is proposed for the MG. The priority order for power supply is wind sources, the hybrid stationary storage system and finally, the power grid connection. The wind generator and the stationary HSS both inject power into the power grid during under-frequency and under-voltage events, and absorb power during over-frequency and over-voltage events, in addition to supplying the loads.

– A fuzzy logic-based droop control system is implemented to regulate the exchange of reactive and active power between the MG and the power grid via an inverter. This control allows for automatic adjustment of the phase and amplitude of the power grid voltage at the point of common coupling (PCC) in both connected and islanded modes of operation for the MG.

4.2. Description of the renewable distributed generator

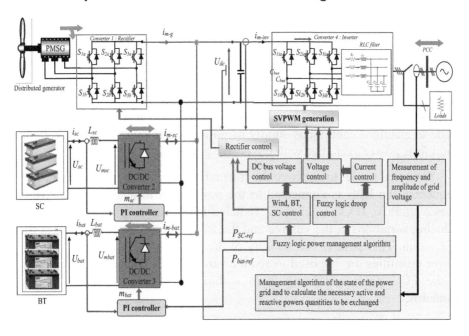

Figure 4.1. *Architecture of the studied distributed generator*

Figure 4.1 illustrates the decentralized renewable generator developed in this chapter. In this hybrid power system, a wind turbine serves as the primary power source, while a battery acts as a long-term energy storage device to reduce wind power peaks by storing excess energy for later discharge during periods of low wind kinetic power. Additionally, a smoothing SC is used to smooth out transient fluctuations in wind power due to rapid changes in wind profile and changes in load and grid demand. The combination of the wind generator and storage systems forms an RPG, which is connected to three-phase loads and the power grid to form an MG. To control the system and meet the power demand of the loads while providing ancillary services to the grid, four different converters are used (a rectifier for a Permanent Magnet Synchronous Generator (PMSG) driven by the wind turbine, a chopper for the battery, a chopper for the SC and a grid connection inverter). The connection of the inverter to the electrical network and loads is facilitated by an RLC filter, which reduces harmonics caused by the switching of the inverter switches and creates a three-phase voltage source at the load terminals.

The MG can work in the following modes:

– Islanded mode: this can be accidental or intentional. This mode is detected in case of a grid fault, where the grid frequency and voltage fluctuations at the PCC point are more severe than the maximum margins mentioned in the standards. In this case, the RPG operates in a disconnected mode to ensure a continuous supply of power to the loads during the fault. Therefore, power exchange with the power grid is stopped.

– Grid-tied mode: to improve the fluctuation plans of the grid voltage and frequency, the RPG is connected to the power grid through a static switch known as PCC (PCC is closed). It helps to maintain system services by reducing variations in the voltage and the frequency of the power grid at the PCC point. This ensures the stability of the power grid. Active and reactive power quantities are imposed to reduce variations in voltage and frequency at the PCC point and bring them within the margins specified in the standards.

4.3. Control of the distributed generator

These are controls for wind turbines, PMSG-side converters, HSS and the DC bus voltage.

4.3.1. *Control of the wind generator*

The wind turbine is controlled to extract the maximum power from the wind. According to Betz's theory, the maximum extractable power from a wind turbine is 59.3% of the available wind power (Kraiem and Abbes 2023). The curve of the power coefficient (C_p) as a function of the speed ratio (λ) for different values of the blade orientation angle (β) is represented in Figure 4.2. This figure shows an optimal value of λ (λ_{opt} = 8.15), corresponding to a maximum value of the power coefficient (C_{pmax} = 0.4794). Adjusting λ to its optimal value with a power coefficient C_{pmax} ensures the extraction of the maximum power (Kraiem and Abbes 2023).

Equation [4.1] gives the expression of the maximum power modulated using the Maximum Power Point Tracking (MPPT) technique.

$$\begin{cases} P_{MPPT} = T_{em-MPPT} \Omega_m \\ T_{em-MPPT} = \frac{1}{2} \frac{\rho \pi R^5 C_{p\max} \Omega_m^2}{\lambda_{opt}^3} \end{cases} \quad [4.1]$$

where the speed ratio λ is given by the following equation:

$$\lambda = \frac{R \Omega_m}{V_v} \quad [4.2]$$

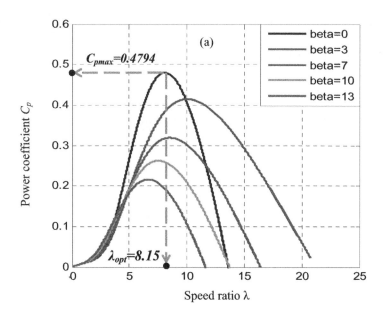

Figure 4.2. *Wind turbine characteristics*

In high winds, it is unavoidable to limit the rotor speed in order to avoid damage to the turbine and the generator. This limitation is achieved by controlling the orientation angle β (Kraiem and Abbes 2023).

The vector control strategy applied to the MPSG consists of imposing a zero direct reference current, i_{sd}*. As a result, the electromagnetic torque, T_{em}, is solely dependent on the quadrature component of the stator current. Thus, the torque and the current i_{sq}* are directly proportional (Hermassi et al. 2022).

$$\begin{cases} i_{sd}^* = 0 \\ i_{sd}^* = \dfrac{T_{em-MPPT}}{p\varphi_m} \end{cases} \quad [4.3]$$

As illustrated in Figure 4.3, the design of the torque control by orientation of the stator current of the PMSG is structured around two proportional integral (PI) type controllers for the currents i_{sd} and i_{sq}. As the transfer functions on the two axes d and q are identical, the integral and proportional parameters of each controller will be equal.

The pitch control is a crucial technique used to regulate the rotational speed of wind turbines. It is activated when the rotor speed exceeds the turbine's maximum rotor speed, denoted as Ω_m, by initiating an increase in the pitch angle of the blades. In the case of variable-speed wind turbines, a mechanical actuator is commonly used to adjust the pitch angle of the blades, which in turn reduces the power coefficient C_p and keeps the speed at its rated value Ω_{mn}. The control strategy is typically implemented as follows:

$$\begin{cases} \beta_{ref} = \beta_0 = 0 \text{ for } 0 \leq \Omega_m \leq \Omega_{mn} \\ \beta_{ref} = \dfrac{\Delta\beta}{\Delta\Omega}(\Omega_m - \Omega_{mn}) + \beta_0 \text{ for } \Omega_m \geq \Omega_{mn} \end{cases} \quad [4.4]$$

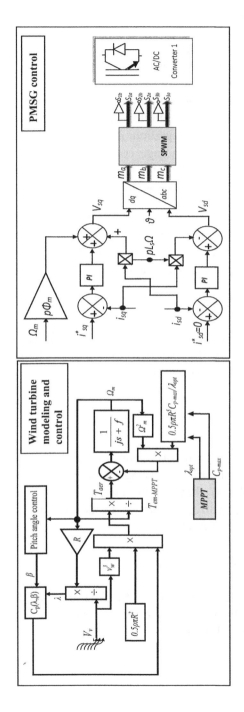

Figure 4.3. Control schemes of the PMSG and the wind turbine

4.3.2. Control of the hybrid storage system

The integration of the HSS enables us to tackle the challenges of incorporating wind power systems into the power grid. Consequently, assessing the value of the storage system is vital to make this extra investment remunerative. To this end, combining two different storage technologies and devising a power management algorithm facilitate the overcoming of the uncertainties posed by the integration of wind energy production, and increases the durability of the storage components. This storage station is based on the combination of two distinct technologies, a storage system with high specific power (SC) and another with high specific energy (battery (BT)).

As shown in Figure 4.4(b), a PI controller is used to adjust the charging/discharging current of the SC to its reference value, $i_{sc\text{-}ref}$. Positive current corresponds to a charging phase of the SC, while negative current corresponds to a discharging phase. The PI controller is defined as follows:

$$U_{msc} = V_{msc} - PI(i_{sc-ref} - i_{sc}) \quad [4.5]$$

The SC DC/DC converter enables the adaptation of the SC voltage to the input voltage of the AC/DC inverter that is necessary. The DC/DC converter 2 is regulated by the duty cycle as follows:

$$m_{sc} = \frac{U_{msc}}{U_{DC}} \quad [4.6]$$

The BT is connected to the DC bus through a bidirectional DC/DC converter (converter 3) in order to manage the power flow for charging and discharging. Figure 4.4(a) shows a PI regulator being used to adjust the charging or discharging current of the BT to its reference value, $i_{bat\text{-}ref}$. The PI controller is defined as follows:

$$U_{mbat} = V_{bat} - PI(i_{bat-ref} - i_{bat}) \quad [4.7]$$

The DC/DC converter of the BT can be used to adjust the BT voltage to the necessary input voltage of the inverter. The duty cycle of the converter 3 control signal is presented in [4.8]:

$$m_{bat} = \frac{U_{mbat}}{U_{DC}} \quad [4.8]$$

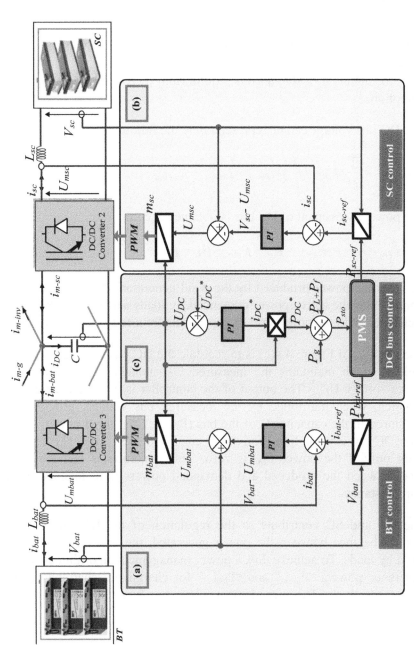

Figure 4.4. Control of the hybrid storage system: (a) control of the BT, (b) control of the SC and (c) control of the DC bus voltage

4.3.3. *Control of the DC bus voltage*

The DC bus voltage fluctuates based on the power exchanged between the wind generator, load and distribution network. As such, a wind generator alone cannot guarantee the regulation of the DC bus voltage. The HSS is utilized to maintain instantaneous equilibrium between production and consumption.

The DC bus voltage can be represented by the following equation:

$$U_{DC} = \frac{1}{C}\int i_{DC} = \frac{1}{C}\int \left(i_{m-g} - i_{m-sc} - i_{m-bat} - i_{m-inv}\right) \qquad [4.9]$$

The power assessment of DC bus is obtained according to:

$$U_{DC} \times i_{DC} = P_{DC} = P_g - P_{sto} - P_L - P_f \qquad [4.10]$$

where P_g is the power produced by the wind generator, P_{sto} is the power to be stored, P_L is the active power demanded by loads and P_f is the power to be exchanged with the power grid to improve its frequency variations plans.

The purpose of Figure 4.4(c) is to regulate the DC bus voltage using a PI controller, which maintains the measured U_{DC} voltage equivalent to the reference voltage U_{DC}^*. The output of the controller is the reference current i_{DC}^*, which, when multiplied by the measured bus voltage, provides the power stored in the capacitor C of the bus (P_{DC}^*). This stored power is then used for DC bus control. In a steady-state condition, the voltage U_{DC} remains constant only if the power P_{DC} is zero. This condition is met when the balance between the produced and demanded powers is maintained under any circumstance.

The BT and SC contribute to the regulation of the DC bus voltage, ensuring a balance between the power generated, the grid power and that required by loads. To achieve this, a power management algorithm estimates the reference powers "P_{sc-ref}" and "$P_{bat-ref}$" for charging/discharging the SC and BT, respectively. This management is based on fuzzy logic technology.

Control of a Wind Distributed Generator 131

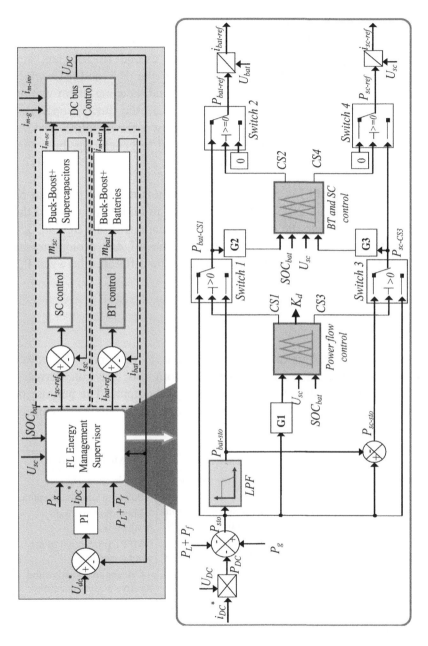

Figure 4.5. Fuzzy logic power management algorithm

4.4. Power management algorithm

The fuzzy logic power management developed in this work has two main objectives: controlling the power flow and minimizing the number of charge/discharge cycles of the BT in order to extend its lifespan, and selecting the appropriate reference powers for the BT and SC while controlling the SOC of each storage system and keeping them within acceptable margins. As illustrated in Figure 4.5, the fuzzy supervisor consists of two blocks: the first is dedicated to "power flow control" and the second, named "BT and SC control", is dedicated to generating the actual reference powers for each storage system.

A low-pass filter (LPF) is employed to filter the amount of power to be stored (P_{sto}) (Abbes et al. 2015). This LPF serves to build the power of the BT "$P_{bat-sto}$" and moderate the drastic power fluctuations in the SC "P_{sc-sto}", thus diminishing the charge/discharge cycle of the BT and prolonging its lifespan. Then, the power P_{sc-sto} is determined by the difference between P_{sto} and $P_{bat-sto}$.

4.4.1. Specifications

The first step in designing the supervision is to determine the power management specifications for the system. From this, we can derive a supervision structure with the necessary inputs and outputs. The purpose of this step is to elucidate the supervisor's objectives, constraints and means of action, as outlined in Table 4.1.

Objectives	Constraints	Means of action
– Improve the variation plan of the frequency of the power grid. – Improve the variation plan of the power grid voltage. – Protect the HSS against the overcharge and the access of discharge. – Reduce the charge/discharge cycle of the battery.	– SOC of the storage system. – Loads and power grid uncertainties. – Insufficiency of the power of wind generator and storage system.	– Power of the battery. – Power of super-capacitors. – Limitation of wind production. – MPPT production of the wind generator. – Calculation of active and reactive powers quantities.

Table 4.1. *Operating specifications of the fuzzy logic power management*

4.4.2. Determination of inputs/outputs

4.4.2.1. Inputs–outputs of the "power flow control" block

Power flow control by fuzzy logic (fuzzy block "power flow control") consists of three inputs and three outputs, as shown in Figure 4.5. The inputs are the difference between the power generated by the wind turbine and the required power ($P_{sto} = P_g - P_L - P_f$), the SOC of the battery (SOC_{bat}) and the SOC of the SC (U_{sc}). The outputs are the switch 1 control signal "CS1", the switch 3 control signal "CS3" and the wind generation degradation factor "Kd" in the event of charging of the HSS. The purpose of this fuzzy toolbox is to deflect rapid power fluctuations in the SC and establish an energy balance.

4.4.2.2. Inputs–outputs of the "BT and SC control" block

In this part, we are going to determine the exact reference powers of the BT and the SC also using fuzzy logic (Fuzzy block "BT and SC control"). This block has four inputs and two outputs, as shown in Figure 4.5. The inputs are $P_{sc\text{-}CS1}$, SOC_{bat}, U_{sc} and $P_{sc\text{-}CS3}$. The outputs are the control signals for switches 2 and 4 (CS2 and CS4) to determine the exact reference powers of the SC and the BT, while respecting the charge and discharge limits. The purpose of this fuzzy toolbox is to keep the SOCs of the SC and the BT at acceptable levels to protect them against overcharging and over-discharging.

In this section, we will use fuzzy logic (fuzzy block "BT and SC control") to determine the exact reference powers of both the BT and SC. This block has four inputs ($P_{sc\text{-}CS1}$, SOC_{bat}, U_{sc}, $P_{sc\text{-}CS3}$) and two outputs (control signals for switches 2 and 4, CS2 and CS4) to ensure the reference powers of both the SC and the BT are within their charge and discharge limits. This fuzzy toolbox aims to maintain acceptable SOC levels for both the SC and the BT, thereby protecting them against overcharging and over-discharging.

4.4.3. Determination of membership functions

This step of the supervisor construction methodology involves determining the membership functions (MFs) that correspond to the inputs and outputs.

4.4.3.1. MFs of inputs–outputs of the "power flow control" block

The determination of the MFs for the fuzzification of the input and output variables of the energy supervisor is an important step in the development of the fuzzy supervisor. As can be seen in Figure 4.6, the MFs for the three input variables and the three output variables of the "power flow control" block have been defined.

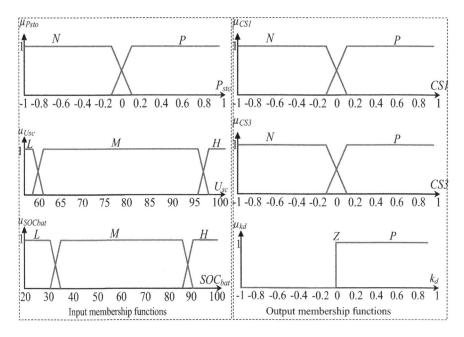

Figure 4.6. *Block diagram of proposed input/output fuzzy-logic-based power flow control*

For the SOC of each storage system (SOC_{bat} and U_{sc}), the MFs are composed of three levels ("L", "M" and "H"), which correspond to the three operating modes. The "L" and "H" sets ensure storage availability, while avoiding low and high levels of saturation. The "M" set is used to make up for missing power, and to store any excess wind energy production.

The values of the power difference P_{sto} can be either "N" or "P", where "N" stands for the discharge of the storage system and "P" represents its charge.

The MF of each power flow management block output (switch 1 control signal (CS1), switch 3 control signal (CS3)) consists of two levels: N means negative and P means positive. The MF of the degradation factor "K_d", as shown in Figure 4.5, considers two fuzzy sets: "Z" indicates that the wind turbine operates in MPPT to produce the maximum power, and "P" indicates that the wind turbine operates without MPPT to only provide the requested power in order to ensure the safe operation of the HSS.

4.4.3.2. MFs of inputs–outputs of the "BT and SC control" block

Figure 4.7 illustrates the MFs for the four input variables and two output variables of the "BT and SC control" block. The MFs of $P_{bat-CS1}$ and P_{sc-CS3} are based on two levels, N and P, to accommodate the requirements of the proposed strategy, with N representing the discharge of the HSS and P representing its charge. The output MFs show the control of switch 2 by control signal CS2 and the control of switch 4 by control signal CS4. The MFs of the control signals are based on two levels: N for negative and P for positive.

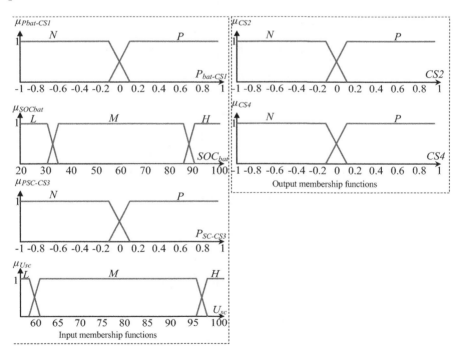

Figure 4.7. Block diagram of proposed input/output fuzzy-logic-based SOC control

4.4.4. Inference engine for energy management

To accurately select the reference powers of the BT and the SC while considering the SOC of each storage system, four switches are utilized. Fuzzy logic is employed to control these switches using the command signals CS1, CS2, CS3 and CS4. Our expertise in the system's operating principle makes it easy to write the fuzzy rules for the developed power supervisor.

4.4.4.1. Fuzzy rules for the "power flow control" block

The inference matrices used to calculate the fuzzy rules of the "power flow control" block are outlined in Tables 4.2–4.4. These tables present the control rules to link the fuzzy inputs and outputs. Table 4.2 displays the rules between P_{sto} and U_{sc} to determine the control signal of switch 1 (CS1). Table 4.3 reveals the rules between P_{sto} and SOC_{bat} to determine the control signal of switch 3 (CS3). Table 4.4 demonstrates the rules between P_{sto} and the SOC of the HSS to enable or disable the K_d degradation factor.

CS1			U_{sc}	
		L	M	H
P_{sto}	P	N	N	P
	N	P	N	N

Table 4.2. CS1 rules

CS3			SOC_{bat}	
		L	M	H
P_{sto}	P	N	N	P
	N	P	N	N

Table 4.3. CS3 rules

K_d			SOC of the HSS	
		L	M	H
P_{sto}	P	Z	Z	P
	N	Z	Z	Z

Table 4.4. K_d rules

Switches 1 and 3 are responsible for improving system performance, in terms of the dynamic behavior of the HSS and their lifespan. Using available production and the SOC of the storage system, the fuzzy supervisor determines the control signals for these two switches:

– Switch 1 allows selecting between P_{sto} and $P_{bat\text{-}sto}$:

  - if P_{sto} is positive and $U_{sc} \geq U_{scmax}$, then $P_{bat\text{-}CS1} = P_{sto}$ else $P_{bat\text{-}CS1} = P_{bat\text{-}sto}$;
  - if P_{sto} is negative and $U_{sc} \leq U_{scmin}$, then $P_{bat\text{-}CS1} = P_{sto}$ else $P_{bat\text{-}CS1} = P_{bat\text{-}sto}$.

– Switch 3 permits choosing between P_{sto} and $P_{sc\text{-}sto}$:

  - if P_{sto} is positive and $SOC_{bat} \geq 90\%$, then $P_{sc\text{-}CS3} = P_{sto}$ else $P_{sc\text{-}CS3} = P_{sc\text{-}sto}$;
  - if P_{sto} is negative and $SOC_{bat} \leq 30\%$, then $P_{sc\text{-}CS3} = P_{sto}$ else $P_{sc\text{-}CS3} = P_{sc\text{-}sto}$.

4.4.4.2. Fuzzy rules for the "BT and SC control" block

The inference matrices used to determine the fuzzy rules of the "BT and SC control" block are described by Tables 4.5 and 4.6. They present the control rules allowing us to associate the inputs and the outputs of the fuzzy supervisor. In fact, Table 4.5 presents the rules between $P_{bat\text{-}CS1}$ and SOC_{bat} to determine the control signal CS2 of switch 2 "CS2". Table 4.6 presents the rules between $P_{sc\text{-}CS3}$ and U_{sc} to determine the control signal CS3 of switch 4.

CS2			SOC_{bat}		
			L	M	H
$P_{bat\text{-}CS1}$	P		N	N	P
	N		P	N	N

Table 4.5. CS2 rules

CS4			U_{sc}		
			L	M	H
$P_{bat\text{-}CS3}$	P		N	N	P
	N		P	N	N

Table 4.6. CS4 rules

To ascertain the BT and SC reference powers, the SOC of each storage system must be taken into account. Switches 2 and 4 are utilized to obtain the specific reference powers for SC and BT, ensuring that the SOC of each storage system remains in acceptable limits, thus safeguarding them from overcharging and over-discharging. These switches are regulated by the fuzzy control signals SC2 and SC4.

– Switch 2 enables choosing between 0 and $P_{bat\text{-}CS1}$:

- if $P_{bat\text{-}CS1}$ is positive and $SOC_{bat} \geq SOC_{bat\text{-}max}$, then $P_{bat\text{-}ref} = 0$ else $P_{bat\text{-}ref} = P_{bat\text{-}CS1}$;

- if $P_{bat\text{-}CS1}$ is negative and $SOC_{bat} \leq SOC_{bat\text{-}min}$, then $P_{bat\text{-}ref} = 0$ else $P_{bat\text{-}ref} = P_{bat\text{-}CS1}$.

– Switch 4 allows selecting between 0 and $P_{sc\text{-}CS3}$:

- if $P_{sc\text{-}CS3}$ is positive and $U_{sc} \geq U_{scmax}$, then $P_{sc\text{-}ref} = 0$ else $P_{sc\text{-}ref} = P_{sc\text{-}CS3}$;

- if $P_{sc\text{-}CS3}$ is negative and $U_{sc} \leq U_{scmin}$, then $P_{sc\text{-}ref} = 0$ else $P_{sc\text{-}ref} = P_{sc\text{-}CS3}$.

Indeed, the low rate of the BT charge is $SOC_{bat\text{-}min} = 30\%$ and the high rate of the BT charge is $SOC_{bat\text{-}max} = 90\%$. Also, the low rate of the SC charge is $U_{scmin} = 58$ V and the high rate of the SC charge is $U_{scmax} = 98$ V.

4.5. Detection and control of the grid faults

For active generators connected to the power grid or to three-phase loads, the most critical aspect in terms of control is the management of the three-phase voltage inverter that facilitates this connection. The inverter control consists of three loops: the droop control loop, which regulates the frequency and voltage at the PCC point; the voltage adjustment loop, which adjusts the voltage at the RLC filter capacitors terminals based on the droop control; and the regulation loop for the currents exchanged with the power grid and loads. These loops are illustrated in Figure 4.8.

In the inverter output, the active and reactive powers transferred to the loads and the grid can be expressed as follows (Vasquez et al. 2009).

$$\begin{cases} P = \dfrac{V_c^2}{R_g^2 + X^2}\Big(R_g\big(V_c - (V_{pcc} - \cos(\delta))\big) + X\, V_{pcc}\sin(\delta)\Big) \\ Q = \dfrac{V_c^2}{R_g^2 + X^2}\Big(X\,\big(V_c - (V_{pcc} - \cos(\delta))\big) - R_g V_{pcc}\sin(\delta)\Big) \end{cases} \quad [4.11]$$

where δ is the transformation angle, which is the phase difference between the inverter output voltage and the grid voltage, V_c is the inverter output voltage, V_{pcc} is the voltage at the point PCC, and $Z = R + jX$ is the impedance of the power grid.

Powers crossing the line depend on line reactance, angles and the levels of the voltage. After using the transformation matrix T, the reactive and active powers become:

$$\begin{pmatrix} P' \\ Q' \end{pmatrix} = T \begin{pmatrix} P \\ Q \end{pmatrix} \qquad [4.12]$$

with: $T = \begin{pmatrix} \dfrac{X}{Z} & -\dfrac{R_g}{Z} \\ \dfrac{R_g}{Z} & \dfrac{X}{Z} \end{pmatrix}$

$$\begin{cases} P' = \dfrac{V_c V_{pcc}}{Z} sin(\delta) \\ Q' = \dfrac{V_c}{Z}\left(V_c - V_{pcc} cos(\delta)\right) \end{cases} \qquad [4.13]$$

If δ is small, the active power can be expressed only as function of the phase and the reactive power is expressed only as function of the voltage as follows:

$$\begin{cases} P' = \dfrac{V_c V_{pcc}}{Z} \delta \\ Q' = \dfrac{V_c}{Z}\left(V_c - V_{pcc}\right) \end{cases} \qquad [4.14]$$

In this section, a droop control system based on fuzzy logic is presented to manage the active and reactive powers (P_f and Q_v) injected or absorbed from the power grid with high precision to ensure its stability based on the outputs of the power management algorithm. The role of the droop control system is to detect the stability state of the power grid and generate adjustment variations that will be subtracted from the voltage phase and amplitude at the PCC point to keep them within acceptable margins outlined in the standards. The control system is composed of two blocks: the fuzzy detection block for identifying the islanded operation mode based on power grid stability state (fuzzy logic islanding detection (FLID)) and the adaptive droop control block based on fuzzy logic (fuzzy logic droop control (FLDC)).

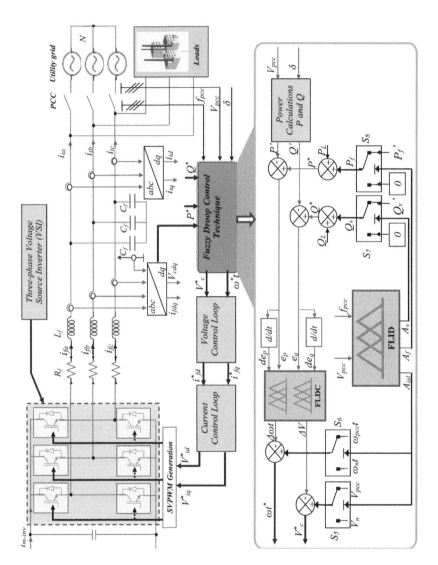

Figure 4.8. Control scheme of the grid-side converter (converter 4)

The objectives, means of action and constraints of the droop control system for power grid stability are summarized in Table 4.7.

Objectives	Constraints	Means of action
– Detect the islanded mode in case of major deviation of grid frequency and voltage. – Improved frequency variation plans. – Improved voltage variation plans.	– Availability of wind generation. – Storage system fully charged/discharged. – Severe change in frequency. – Severe voltage variation.	– Calculation of the necessary active power to adjust the frequency. – Calculation of the necessary reactive power to adjust the voltage. – Control of the storage system.

Table 4.7. *Operating specifications of the FLID*

4.5.1. *Fuzzy logic islanding detection*

The FLID performs the following roles:

– The droop control system detects the standalone operation mode when the frequency and voltage variations at the PCC exceed the maximum margins outlined in the standards. During this mode, the RPG disconnects from the power grid and only supplies the installed loads until the issue is resolved.

– If the voltage and frequency variations are within the maximum margins, the FLID detects the grid-connected operation mode of the RPG. However, repetitive variations within the standards can damage electronic devices. In this scenario, the FLID reduces these variations by switching S7 and S8 to impose the necessary quantities of active and reactive power (P_f and Q_v) to be exchanged with the power grid. This ensures power grid stability by improving the frequency and voltage magnitude variations at PCC. The power quantities are calculated as follows:

$$\begin{cases} P_f = k_f \left(\Delta f - \text{sign}(\Delta f) \, \Delta f_{acc} \right) \\ Q_v = k_v \left(\Delta V - \text{sign}(\Delta V) \, \Delta V_{acc} \right) \end{cases} \quad [4.15]$$

where k_f and k_v are adjustable gains, $\Delta f = f_{pcc} - f_n$ is the difference between the grid and nominal frequencies, $\Delta V = V_{pcc} - V_n$ is the difference between the grid and nominal voltages, $\Delta f_{acc} = \pm 0.1$ Hz is the acceptable frequency level, and $\Delta V_{acc} = \pm 5$ V is the acceptable voltage level.

To meet the previously stated stability objectives, two inputs and three outputs are proposed for the fuzzy toolbox FLID. The inputs are the grid voltage and frequency V_{pcc} and f_{pcc} at the PCC. The outputs are the signals (A_{isl}, A_v and A_f) dedicated to controlling automatic switches S5, S6, S7 and S8.

The input and output MFs of this FLID ensure the transition between different operation modes, as shown in Figure 4.9.

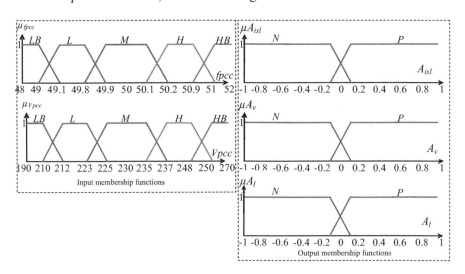

Figure 4.9. Block diagram of proposed FLID inputs/outputs

The input MFs are selected symmetrically, but for each input, we use five fuzzy sets: LB (low big), L (low), M (medium), H (high), and HB (high big). Similarly, the output variables' MFs are symmetrical, but we use two fuzzy sets: N for negative and P for positive.

The control rules that associate inputs with outputs for islanding detection are obtained through the analysis of frequency and voltage grid fluctuations using fuzzy logic. The control rules associating inputs with outputs are composed as follows:

– Table 4.8 shows the fuzzy rules used between V_{pcc} and f_{pcc} to determine the control signal A_{isl} of switches 5 and 6. There are 20 rules.

– Table 4.9 presents the fuzzy rules used for V_{pcc} to determine the signal of A_v command of switch 7. There are five rules.

– Table 4.10 gives the fuzzy rules used for f_{pcc} to determine the signal of A_f command of switch 8. There are five rules.

A_{isl}			V_{pcc}			
		HB	H	M	L	LB
	HB	N	N	N	N	N
	H	N	P	P	P	N
f_{pcc}	M	N	P	P	P	N
	L	N	P	P	P	N
	LB	N	N	N	N	N

Table 4.8. A_{isl} rules

			V_{pcc}			
		HB	H	M	L	LB
A_v		N	P	N	P	N

Table 4.9. A_v rules

			f_{pcc}			
		HB	H	M	L	LB
A_f		N	P	N	P	N

Table 4.10. A_f rules

Switches S5 and S6 allow for the selection between V_n and V_{pcc}, and f_n and f_{pcc}, respectively. If A_{isl} is positive, then S5 and S6 are in the V_{pcc} and f_{pcc} states, respectively. Otherwise, S5 and S6 are in the V_n and f_n states, respectively. Switch S7 enables the choice between 0 and Q_v'. If A_v is positive, then $Q_v = Q_v'$; otherwise, $Q_v = 0$. Switch S8 allows the selection between 0 and P_f'. If A_f is positive, then $P_f = P_f'$; otherwise, $P_f = 0$.

In grid-connected mode, the automatic switches S5 and S6 are, respectively, in the states V_{pcc} and $\omega_{pcc}t$, while S7 and S8 are, respectively, in the states Q_v' and P_f'. The droop control system reduces voltage and frequency fluctuations by adjusting the power supplied by the inverter to the total power consumed by the loads and the grid, which is represented by $(P^* = P_L + P_f)$ and $(Q^* = Q_L + Q_v)$. This system controls the powers exported or imported (P_f and Q_v) to or from the power grid to ensure its stability.

In standalone mode, once FLID detects an isolation condition, the renewable generator will switch to standalone mode and any additional injection of (P_f and Q_v) will be irrelevant, resulting in zero values for P_f and Q_v in this mode. Automatic switches S7 and S8 will switch to state 0, while S5 and S6 will switch to states V_n and $\omega_n t$.

4.5.2. *Fuzzy droop control technique for the adjustment of the grid frequency and voltage*

The aim of using fuzzy logic in this section is to generate adjustment variations that will be incorporated into the phase and amplitude of the grid voltage to keep them within better margins than those specified in the standards. The FLDC used in this study has four inputs and two outputs. The four inputs are the errors ($e_p = P^* - P'$) and ($e_q = Q^* - Q'$), as well as the derivatives of the errors de_q and de_p. The outputs are the variations of the phase and the amplitude of the power grid output voltage, $\Delta\omega t$ and ΔV. The input and output MFs of the FLDC are illustrated in Figure 4.10.

The MFs of the errors have a symmetric shape. To account for the sensitivity of the variables to be controlled, nine subsets are introduced: NBB (negative big big), NB (negative big), NM (negative medium), NS (negative small), Z (zero), PBB (positive big big), PB (positive big), PM (positive medium), and PS (positive small). The MFs of the error derivatives "dep and deq" are selected to specify the response of the system. Therefore, three fuzzy subsets are introduced: P (positive), M (medium) and N (negative).

The MFs of the output variables also have a symmetric shape. To accurately describe the frequency and voltage variation errors, nine fuzzy subsets are introduced.

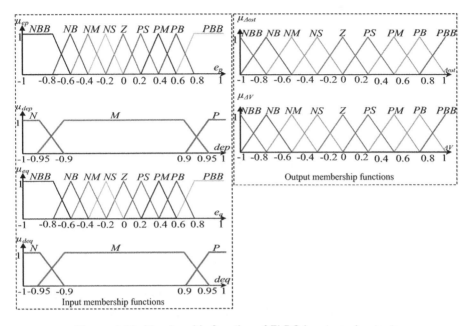

Figure 4.10. *Membership function of FLDC inputs and outputs*

The next step involves developing droop control rules. Based on the behavior of the process and the action of the command variation to be applied, tables of fuzzy rules are deduced. The frequency and voltage control responses are summarized in Tables 4.11 and 4.12.

$\Delta\omega t$			e_p						
		NBB	NB	NM	NS	Z	PS	PM	PB	PBB
	N	PS	PS	PS	PS	PS	PS	PS	PS	PS
de_p	M	PBB	PB	PM	PS	Z	NS	NM	NB	NBB
	P	NS	NS	NS	NS	NS	NS	NS	NS	NS

Table 4.11. *Rules for frequency control*

ΔV			e_q							
			NBB	NB	NM	NS	Z	PS	PM	PB	PBB
de_q	N	PS	PS	PS	PS	PS	PS	PS	PS	PS
	M	PBB	PB	PM	PS	Z	NS	NM	NB	NBB
	P	NS	NS	NS	NS	NS	NS	NS	NS	NS

Table 4.12. *Rules for voltage control*

This set of rules accounts for all potential scenarios within the evaluated system, encompassing various error values and their associated fluctuations. In the study, fuzzy compensators were implemented to control the transfer of active and reactive powers from the inverter to the power grid, accomplished by automatic adjustments to the voltage amplitude and frequency at the PCC point.

4.6. Simulation results

This section will detail the simulation results of the control strategy designed to maintain power grid stability in the face of fluctuations in wind profile, loads and grid voltage frequency and magnitude. This simulation was conducted using the MATLAB/SIMULINK environment.

In order to evaluate the control technique applied to the renewable generator, the test profiles are shown in these figures:

– The wind speed profile presented in Figure 4.11 over a period of 72 s is applied.

– Figure 4.12 shows the active power profile demanded by loads.

– Figures 4.13(a) and (b) represent the profiles of the voltage and the frequency of the power grid before regulation, respectively.

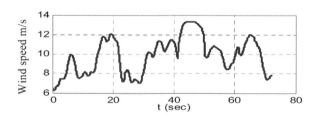

Figure 4.11. *Wind profile*

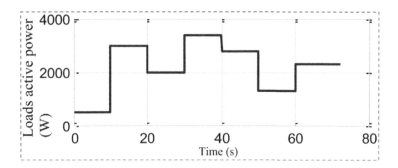

Figure 4.12. *Power demanded by loads*

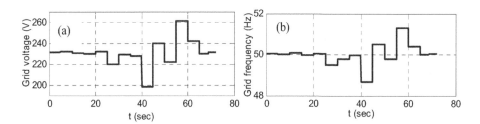

Figure 4.13. *Power grid characteristics before regulation:
(a) grid voltage V_{pcc} and (b) grid frequency f_{pcc}*

4.6.1. Control and power management of the distributed generator

According to the internal specifications of the two storage devices presented in this study, an LPF is used with a constant time $\tau = 6$ s (Abbes et al. 2015). Figure 4.14 illustrates the evolution of the power to be stored to satisfy the consumption scenario.

Figures 4.15 and 4.16 depict the response of two storage technologies, BT and SC. The BT technology has been designed for a medium storage system, and aims to extend its lifetime by reducing the number of charge/discharge cycles, whereas the SC is used for fast response to the power balance between the wind generator and the power grid and loads.

Figures 4.15(a) and (b) illustrate the BT charge/discharge reference power obtained using both the fuzzy logic power management algorithm and the rule-based algorithm described in our previous publication (Krim et al. 2019). Likewise, for the SC, Figures 4.16(a) and (b), respectively, display the charge/discharge reference power obtained using classical control and fuzzy logic.

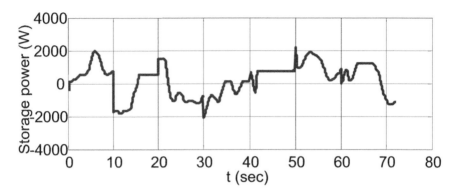

Figure 4.14. *Profile of the power to be stored (difference between production and demand)*

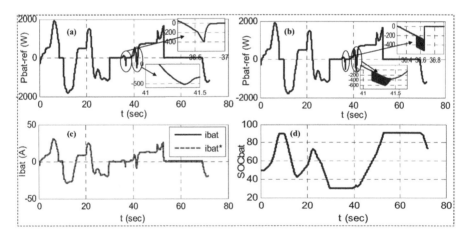

Figure 4.15. *Simulation results for the BT: (a) BT reference power with fuzzy logic, (b) BT reference power with rule-based algorithm, (c) BT charge/discharge current and (d) BT state of charge*

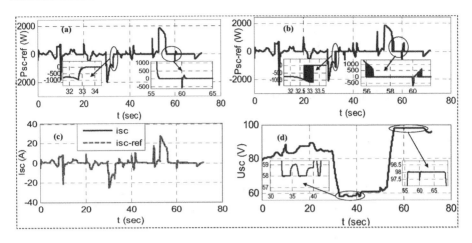

Figure 4.16. *Simulation results for the SC: (a) SC reference power with fuzzy logic, (b) SC reference power with rule-based algorithm, (c) SC charge/discharge current and (d) SC state of charge*

Figures 4.15(c) and 4.16(c) show the charge/discharge currents for the BT and the SC, respectively. The BT responds slowly to power demands, while the SC provides transient currents. To prevent excessive discharge, when the SC reaches its minimum charge level (U_{scmin}), the BT provides the remaining power demand, and vice versa when the BT reaches its minimum SOC, the SC provides the necessary power. If the SC reaches U_{scmax}, charging is stopped, and the BT absorbs the excess power to avoid overcharging. Similarly, when the BT reaches SOC_{max}, charging is stopped, and the SC absorbs the excess power. These simulation results demonstrate the effectiveness of the proposed fuzzy supervisor in maintaining U_{sc} voltage (Figure 4.15(d)) and SOC_{bat} (Figure 4.16(d)) in acceptable levels.

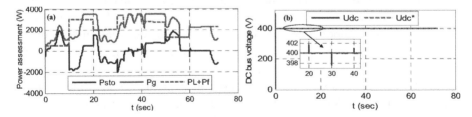

Figure 4.17. *Characteristics at the common DC bus: (a) simulated power assessment at the DC bus; (b) DC bus voltage*

Therefore, the fuzzy logic-based algorithm represents reliable and efficient power management. Thus, the simulation results prove the effectiveness of the proposed strategy by ensuring the balance between production and consumption (Figure 4.17(a)) to maintain a stable DC bus voltage at 400 V (Figure 4.17(b)).

4.6.2. Detection and correction of the grid voltage and frequency variations at the PCC

The distributed generator can operate in either grid-connected or standalone mode, depending on the state of the power grid's frequency and voltage.

Figure 4.18(c) displays the time periods in which the generator operates in standalone mode (A_{isl} = −0.5) or in grid-connected mode (A_{isl} = 0.5). During grid-connected operation, the generator exchanges active and reactive power with the power grid, which helps to mitigate voltage and frequency fluctuations and keep them within more rigorous thresholds than those specified in the standards. These power exchanges are depicted in Figures 4.18(a) and (b), respectively.

In this study, we compare a generalized droop control system, which was presented in one of our previous publications (Krim et al. 2019), with the intelligent droop control technique that is based on fuzzy logic.

Figure 4.19 depicts the grid frequency before and after regulation using the fuzzy and classical control methods detailed in our previously published paper (Krim et al. 2019). The graph shows that the active power "P_f" exchanged with the power grid has a significant impact on reducing frequency fluctuations. These variations are effectively minimized, highlighting the importance of both fuzzy and conventional control techniques in enabling renewable generators to participate in system services. Similarly, as shown in Figure 4.20, the impact of the reactive power "Q_v" exchanged with the power grid on reducing voltage fluctuations is also remarkable. With the fuzzy method, the amplitude and frequency levels of the power grid voltage at the PCC point are well minimized, and fall within the optimal margins of [50.1, 49.9] and [235, 225], respectively.

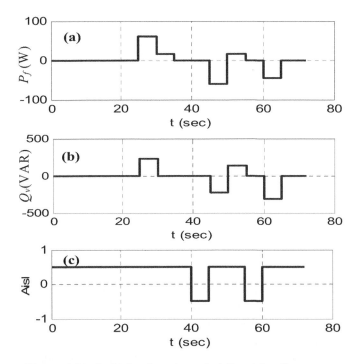

Figure 4.18. *Anti-islanding characteristics: (a) active power exchanged with the power grid; (b) reactive power exchanged with the power grid; (c) anti-islanding signal*

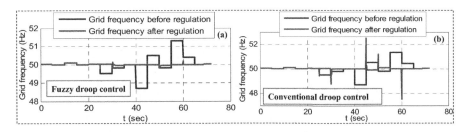

Figure 4.19. *Frequency at PCC before and after regulation: (a) with FLDC; (b) with conventional droop control*

In order to prove the effectiveness of the applied control strategy, the statistical tests summarized in Table 4.13 show that up to a frequency of

50 ± 1 Hz, the control strategy reduces the frequency variation plane and brings it back within the acceptable margin "50 ± 0.1 Hz" and beyond the maximum margin "50 ± 1 Hz", standalone mode is detected. Similarly for the voltage, it is reduced from 230 ± 20 V to 230 ± 5 V.

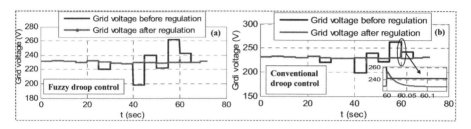

Figure 4.20. *Voltage at PCC before and after regulation: (a) with FLDC; (b) with conventional droop control*

Table 4.13 summarizes the statistical tests used to demonstrate the effectiveness of the applied control strategy. The results show that the control strategy reduces the frequency variation plane and brings it back within the acceptable margin of "50 ± 0.1 Hz" up to a frequency of 50 ± 1 Hz. Beyond the maximum margin of "50 ± 1 Hz", standalone mode is detected. Similarly, the voltage is reduced from 230 ± 20 V to 230 ± 5 V. These results indicate that the control strategy is effective in regulating both frequency and voltage, meeting the acceptable standards within the specified margins.

Time	0–5	5–10	10–15	15–20	20–25	25–30	30–35	35–40
$f_{initial}$	50.02	50	50.07	49.98	50.03	49.5	49.8	49.98
f_{final}	50.02	50	50.07	59.98	50.03	50.02	49.99	49.98
$V_{initial}$	231.5	232	231	230	232	220	229	228
V_{final}	231.5	232	231	230	232	230.5	229	228

Time	40–45	45–50	50–55	55–60	60–65	65–69	69–72
$f_{initial}$	48.68	50.5	49.8	51.3	50.4	50	50.02
f_{final}	50	50	49.98	50	50.01	50	50.02
$V_{initial}$	198	240	222	262	242	230	231.5
V_{final}	230	229.5	229.2	230	230.4	230	231.5

Table 4.13. *Statistical tests of frequency and voltage*

The results indicate that the methodology presented in this study can effectively enhance the energy efficiency of a grid-connected renewable distributed generator that is equipped with an HSS. Furthermore, the findings demonstrate that the proposed control strategy for the distributed generator is feasible in various operational modes, including ancillary services to ensure grid stability during grid faults. By using an intelligent control method, the proposed approach enables optimal frequency and voltage variations within the grid, with significant reductions in disturbances caused by the limited performance of the classical control technique (Krim et al. 2019), which is known for its high overshoot and long response time. A comparison framework of the fuzzy droop control and a classical droop control under the case studies is provided in Table 4.14 using various criteria.

Comparison criteria	Grid frequency f_{pcc} stability				Grid voltage V_{pcc} stability			
	At t = 45s		At t = 60s		At t = 45s		At t = 60s	
	GDC	FLDC	GDC	FLDC	GDC	FLDC	GDC	FLDC
Percentage of overrun	6	0.2	5.8	0.4	5.21	1.7	10.43	1.6
Stabilization time (s)	0.17	0.04	0.2	0.06	0.16	0.035	0.172	0.06
Rising time (s)	0.2	0.11	0.2	0.15	0.2	0.11	0.2	0.15
Control capability	Yes		Yes		Yes		Yes	
Control performance	High peak	Good	High peak	Good	High peak	Good	High peak	Good
Robustness	Poor	Robust	Poor	Robust	Poor	Robust	Poor	Robust

Table 4.14. *Comparative study between proposed control approach and conventional control*

This comparative study demonstrates that the suggested intelligent control strategy is suitable for the RDG application, exhibiting highly robust behavior in adjusting the grid frequency and voltage at PCC.

4.7. Conclusion

This chapter proposes a new smart control technique, based on fuzzy logic, for controlling a decentralized renewable active generator that operates in both connected and autonomous modes, in accordance with the stability state of the electrical network. The control method comprises three main control blocks:

– The first block is dedicated to power management using fuzzy logic, which controls two battery/SC storage technologies. The goal is to meet the needs of the various stakeholders involved in integrating wind energy into the electrical network and manage the exchanges between different energy sources.

– The second block is an intelligent stability control that uses fuzzy logic to regulate the flow of power exchanged at the point of common coupling between the active generator and the electrical network based on the detected mode. The aim of this control is to ensure network stability by maintaining frequency and voltage within optimal margins.

– The third block is a fuzzy detector that receives instantaneous measurements of the frequency and voltage at the PCC point with the power grid. The inputs are compared to the maximum margins specified by standards and control limits to determine the desired operating mode. In the event of a failure (severe variations of frequency and voltage), the autonomous mode is detected to provide continuous power supply to loads and protect equipment and people from severe frequency and voltage fluctuations. If the frequency and voltage fluctuations are within the standards, the connected mode is detected.

The results for various scenarios demonstrate the robustness of fuzzy logic in supporting wind generators to enhance the stability of the electrical network.

4.8. References

Abbes, D., Bensmaine, F., Labrunie, A., Robyns, B. (2015). Energy management and batteries lifespan estimation in a photovoltaic system with hybrid storage: A comparative study. *13th Brazilian Power Electronics Conference and 1st Southern Power Electronics Conference (COBEP/SPEC)*, Fortaleza.

Almaksour, K., Krim, Y., Kouassi, N., Navarro, N., François, B., Letrouvé, T., Saudemont C., Taunay L., Robyns, B. (2021). Comparison of dynamic models for a DC railway electrical network including an AC/DC bi-directional power station. *Mathematics and Computers in Simulation*, 184, 244–266.

Cabrane, Z., Ouassaid, M., Maaroufi, M. (2017). Battery and supercapacitor for photovoltaic energy storage: A fuzzy logic management. *IET Renewable Power Generation*, 1157–1165.

Cheikh-Mohamad, S., Sechilariu, M., Locment, F., Krim, Y. (2021). PV-powered electric vehicle charging stations: Preliminary requirements and feasibility conditions. *Applied Sciences*, 11, 1770.

Hermassi, M., Krim, S., Krim, Y., Hajjaji, M.A., Mtibaa, A., Mimouni, M.F. (2022). Xilinx-FPGA for real-time implementation of vector control strategies for a grid-connected variable-speed wind energy conversion system. *5th International Conference on Advanced Systems and Emergent Technologies (IC_ASET)*, Hammamet.

Kofinas, P., Dounis, A.I., Vouros, G.A. (2018). Fuzzy Q-learning for multi-agent decentralized energy management in microgrids. *Applied Energy*, 219, 53–67.

Kraiem, Y. and Abbes, D. (2023). Modeling, control, and simulation of a variable speed wind energy conversion system connected to the power grid. In *Encyclopedia of Electrical and Electronic Power Engineering*, García, J. (ed.). Elsevier, 3, 485–501. doi: 10.1016/B978-0-12-821204-2.00086-6.

Krim, Y., Abbes, D., Krim, S., Mimouni, M.F. (2018a). Intelligent droop control and power management of active generator for ancillary services under grid instability using fuzzy logic technology. *Control Engineering Practice*, 81, 215–230.

Krim, Y., Abbes, D., Krim, S., Mimouni, M.F. (2018b). Control and fuzzy logic supervision of a wind power system with battery/supercapacitor hybrid energy storage. *7th IEEE International Conference on Systems and Control (ICSC)*, Valencia.

Krim, Y., Abbes, D., Krim, S., Mimouni, M.F. (2018c). Power management and second-order sliding mode control for standalone hybrid wind energy with battery energy storage system. *Proceedings of the Institution of Mechanical Engineers. Part I: Journal of Systems and Control Engineering*, 232, 1389–1411.

Krim, Y., Abbes, D., Krim, S., Mimouni, M.F. (2018d). Fuzzy droop control for voltage source inverter operating in standalone and grid connected modes. *15th IEEE International Multi-Conference on Systems, Signals & Devices (SSD)*, Hammamet.

Krim, Y., Abbes, D., Krim, S., Mimouni, M.F. (2019). A flexible control strategy of a renewable active generator to participate in system services under grid faults. *International Transactions on Electrical Energy Systems*, 29, e2687.

Krim, Y., Abbes, D., Krim, S., Mimouni, M.F. (2020a). STA and SOSM control-based approach of a renewable power generator for adjusting grid frequency and voltage. *International Transactions on Electrical Energy Systems*, 30, e12363.

Krim, Y., Abbes, D., Robyns, B. (2020b). Joint optimization of sizing and fuzzy logic power management of a hybrid storage system considering economic reliability indices. *IET Renewable Power Generation*, 14, 2581–2591.

Krim, Y., Sechilariu, M., Locment, F. (2021). PV benefits assessment for PV-powered charging stations for electric vehicles. *Applied Sciences*, 11, 4127.

Krim, Y., Krim, S., Mimouni, M.F. (2022a). Optimization of an adaptive droop control and powers allocation for a distributed generator under loads and grid uncertainties. *Sustainable Energy, Grids and Networks*, 32, 100950.

Krim, Y., Sechilariu, M., Locment, F., Alchami, A. (2022b). Global cost and carbon impact assessment methodology for electric vehicles PV-powered charging station. *Applied Sciences*, 12, 4115.

Lamnatou, C., Chemisana, D., Cristofari, C. (2022). Smart grids and smart technologies in relation to photovoltaics, storage systems, buildings and the environment. *Renewable Energy*, 185, 1376–1391.

Mendis, N., Muttaqi, K.M., Perera, S. (2014). Management of low- and high-frequency power components in demand-generation fluctuations of a DFIG-based wind-dominated RAPS system using hybrid energy storage. *IEEE Transactions on Industry Applications*, 2258–2268.

Rekik, M., Abdelkafi, A., Krichen, L. (2013). A novel control strategy of a distributed generator operating in seven modes for ancillary services under grid faults. *International Journal of Electrical Power & Energy Systems*, 100–108.

Shafiee, Q., Stefanović, Č., Dragičević, T., Popovski, P., Vasquez, J-C., Guerrero, J.-M. (2014). Robust networked control scheme for distributed secondary control of islanded microgrids. *IEEE Transactions on Industrial Electronics*, 61, 5363–5374.

Taghizadeh, M., Hoseintabar, M., Faiz, J. (2015). Frequency control of isolated WT/PV/SOFC/UC network with new control strategy for improving SOFC dynamic response. *Transactions on Electrical Energy Systems*, 25, 1748–1770.

Vasquez, J.C., Guerrero, J.M., Luna, A., Rodriguez, P., Teodorescu, R. (2009). Adaptive droop control applied to voltage-source inverters operating in grid-connected and islanded modes. *IEEE Transactions on Industrial Electronics*, 56(10), 4088–4096.

Yan, X., Abbes, D., Labrunie, A., Krim, Y., Robyns, B. (2020). Economic analysis of a hybrid storage system associated to PV sources and supervised by fuzzy logic power management. *ELECTRIMACS 2019*, Springer, Cham.

5

Fault-Tolerant Control of Sensors and Actuators Applied to Wind Energy Systems

5.1. Introduction

Nonlinearities and system uncertainties are the most important difficulties in designing controllers that ensure stability and acceptable closed-loop performance. Many significant results on the stability and robust control of uncertain nonlinear systems using the TS fuzzy model have been reported over the past decades, and considerable advances have been made (Connor et al. 1992; Chen et al. 1996; Rocha et al. 2005; Boukhezzar et al. 2006; Kamal et al. 2011, 2012a, 2013; Khanl et al. 2011). However, as stated in Khan and Hossain (2011), many approaches for stability and robust control of uncertain systems are often characterized by conservatism when dealing with uncertainties. It has been well known that the TS fuzzy model is very effective representation of complex nonlinear systems. In the TS fuzzy model, the state space of a nonlinear system is divided into different fuzzy regions with a local linear model being used in each region. The overall model output is obtained by defuzzification using the center of gravity (COG) method. Once the fuzzy model is obtained, control design can be carried out via the so-called parallel and distributed computing (PDC) approach, which employs multiple linear controllers corresponding to the locally linear plant models (Boukhezzar et al. 2006). This class of systems is

---

Chapter written by Elkhatib KAMAL and Abdel AITOUCHE.

For a color version of all figures in this chapter, see www.iste.co.uk/benkhaderbouzid/fault.zip.

described as a weighted sum of some simple linear subsystems, and thus is easily analyzable.

Commonly seen in practice, many control systems are subjected to faults which can be caused by sensors, actuators or systems faults. Therefore, it is an important issue in control system design as to how the system is kept stable and acceptable performance levels maintained when a failure occurs. Generally speaking, FTC can be achieved either passively or actively. In passive FTC, the formed may be viewed as robust control. It requires a priori knowledge of possible faults that may affect the system and its controller is based on treating all possible faults as uncertainties, which are taken into account for the design of tolerant control by using different techniques such as H_∞ (Guo 2010; Patton 1997). The interest of this approach lies in the fact that no online information is required and the structure of the control law remains unchanged. Generally, the structure of the uncertainties (faults) is not taken into account in order to lead to a convex optimization problem. Furthermore, the class of the faults considered is limited and it then becomes risky to use only passive FTC (Niemann and Stoustrup 2005). While in active FTC, the controller is designed to be reconfigurable for occurring faults according to the fault detection and estimation performed by an observer to allow the faulty system to accomplish its mission. Indeed, the active FTC has been introduced to overcome the passive control drawbacks. In addition, the active FTC can usually get better control performances; it has attracted more attention in recent years. The knowledge of some information about these is required and obtained from a fault detection and diagnosis (FDD) block. Different ideas are developed in the literature, for example, control law rescheduling (Leith et al. 1999; Mufeed et al. 2003; Ocampo-Martinez et al. 2010). This approach requires a very robust fault detection and isolation (FDI) block, which constitutes its major disadvantage. Indeed, a false alarm or an undetected fault may lead to degraded performance or even to instability. In the linear framework, the FTC problem is widely treated (Stilwell et al. 1997; Marx et al. 2004; Zhang and Jiang 2008). However, in practice, most of physical systems are nonlinear; hence, it is primary to consider the FTC design for nonlinear systems. Some approaches dealing with this problem are proposed by Zuo et al. (2010) and Shtessel et al. (2002). The proposed methods in Yao et al. (2009) are confined to the actuator failures and did not consider the fuzzy systems with sensor failures. Afterward, several FTC methods have been extended to the nonlinear systems with sensor faults (Tong et al. 2002). In the above studies, the considered faults affecting the system behavior are modeled by a constant function. However, in practice, the faults are often time variant and as we know, in practice, many control systems are subjected to faults, which can be caused by actuators, sensors or systems faults and parameter uncertainties. In addition, the problem of control in the maximization of power generation in VS-WES has been greatly studied and such control still remains an active research area (Mohamed et al. 2001; Datta et al. 2003; Camblong et al. 2006; Koutroulis et al. 2006; Abo-Khalil et al. 2008; Galdi 2009; Masoud et al. 2009; Aggarwal et al. 2010; Whei-Min et al. 2010).

However, in this chapter, our goal is to present three different proposed approaches, namely, robust fuzzy fault-tolerant control (RFFTC), robust fuzzy scheduled fault-tolerant control (RFSFTC) and robust dynamic fuzzy fault-tolerant control (RDFFTC) for TS models subject to time varying actuator faults, sensor faults and parameter uncertainties, and maximize the output power from the wind turbine (WT).

5.2. Objective

WTs are nonlinear mechanical systems exposed to wind profiles that are difficult to control. As such, the control of WTs is difficult due to the lack of systematic methods to guarantee performance subject to parameter variations (change in temperature, pressure, load, aging of components, variation in resistance of a power converter, measurement error, etc.).

Stability is one of the most important issues in the analysis and synthesis of control systems. Since faults are frequently a source of instability and are encountered in various engineering systems, robust fuzzy control design for nonlinear systems in the presence of faults has received considerable interest. The performance of a WT is highly dependent on the sensors and actuators. A sudden failure of any of the sensors or actuators decreases the performance of the system. Furthermore, if a failure is not detected and addressed quickly, its effect will lead to system shutdown. Therefore, to reduce the failure rate and prevent unplanned shutdowns, a real-time fault detection, isolation and compensation method could be adopted.

Therefore, it is important when designing a control system to know how the system remains stable and how acceptable performance levels are maintained when a failure occurs. Generally speaking, fault-tolerant control (FTC) can be achieved passively or actively. In the case of passive FTC, training can be considered as robust control. It requires an a priori knowledge of the possible failures that can affect the system and its controller is based on the treatment of all possible failures as uncertainties that are considered for the design of the tolerant control using different techniques such as H_∞. The advantage of this approach is that no online information is needed and the structure of the control law remains unchanged. Generally, the structure of the uncertainties (faults) is not considered in order to lead to a convex optimization problem. Moreover, the class of faults considered is limited and it then becomes risky to use only passive FTC, whereas in the case of active FTC, the controller is designed to be reconfigurable for faults that occur according to the detection and estimation of faults performed by an observer to allow the faulty system to accomplish its mission.

Indeed, active FTC was introduced to overcome the drawbacks of passive control. FTC is based on a FDD system. Different methods are developed in the literature in the linear or linearized case around an operating point. Most of the methods deal only with sensor or actuator faults but never with their combination.

In the last decade, various methods for fault detection and isolation and fault compensation have been developed. Most of the methods are based on systems linearized around an operating point and consider sensor or actuator faults and the assumption that only one set of sensors or actuators can have faults at a time and do not consider the nonlinearities of the system.

This chapter proposes an observer-based actuator or sensor detection scheme for TS (Tanaka and Sugeno 1992) type fuzzy systems subject to sensor faults, parametric uncertainties and actuator faults. The detection system provides residuals for detecting and isolating sensor faults that may affect a TS model. The fuzzy TS model is adopted for fuzzy modeling of the uncertain nonlinear system and establishing fuzzy state observers. Sufficient conditions are established for robust stabilization in the sense of Lyapunov stability for the fuzzy system. The sufficient conditions are formulated in the form of linear matrix inequalities (LMI). The effectiveness of the proposed controller design method is finally demonstrated on a DFIG-based WT to illustrate the effectiveness of the proposed method.

The chapter is organized as follows. Section 5.2 presented an introduction. The RFFTC control for the WES subject to time varying actuator faults, time varying sensor faults and parameter uncertainties are presented in section 5.3. The RFSFTC control for the WES subject to actuator faults, sensor faults with time varying and parameter uncertainties are presented in section 5.4. Intelligent RDFFTC of WES in the presence of parameter uncertainties, sensor fault and actuator faults are studied in section 5.5. A conclusion will be given in section 5.6.

5.3. RFFTC of WES with DFIG

This section presents a new method for RFFTC of nonlinear systems described by TS fuzzy systems subject to sensor faults, parameter uncertainties and time varying actuator faults. The algorithm based on reconfiguration mechanism is then investigated for detection, isolation and accommodation of faults. The idea is to use fuzzy FDOS (fuzzy dedicated observers), FPIEO and design a new control law to minimize the state deviation between a healthy observer and the eventually faulty actual model. This scheme requires the knowledge of the system states and the occurring faults. These signals are estimated from an FDOS and FPIEO (proportional-integral estimation observer). TS fuzzy systems are classified into three families based on the input matrices and a design RFFTC for each family. In each family, the FTC law is designed by using the Lyapunov method to obtain conditions, which are given in LMIs formulation. The effectiveness of the proposed controller design methodology is finally demonstrated through a wind energy system with DFIG to illustrate the effectiveness of the proposed method.

This section is organized as follows. Section 5.3.1 provides the proposed RFFTC scheme, TS fuzzy model FDOS and FPIEO. The proposed algorithm and the augmented system are presented in section 5.3.2. Section 5.3.3 shows the stability and robustness conditions for the proposed algorithm followed by the calculation of state RFFTC, FDOS and FPIEO gains. Section 5.3.4 shows WES model system with DFIG (doubly fed induction generators (IGs)). Simulation and results are shown in section 5.3.5.

5.3.1. TS fuzzy model with parameter uncertainties and fuzzy observer

In this section, we will present the TS fuzzy plat model subject to sensor faults, parametric uncertainties and time varying actuator faults. In addition, we will present the FDOS and FPIEO and the RFFTC scheme design.

5.3.1.1. TS fuzzy plant model with parameter uncertainties, sensor faults and actuator faults

This work considers the occurrence of actuator and sensor faults. *For the first family*, if we take the non-time varying the parameter uncertainties into consideration, we write the plant dynamics as follows (Tanaka et al. 1992; Chen et al. 2007):

$$\dot{x}(t) = \sum_{i=1}^{p} \mu_i(q(x(t)))[A_i x(t) + \alpha_i Bu(t) + \bar{D}_i f_a(t)] + \sum_{i=1}^{p} \mu_i(q(x(t)))\Delta A_i x(t)$$

$$y(t) = \sum_{i=1}^{p} \mu_i(q(x(t)))[(I + F_s) C_i x(t)] \qquad [5.1]$$

where $\bar{D}_i \in \kappa^{n \times k}$ are known matrices of actuator faults and $f_a(t) \epsilon \kappa^{k \times 1}$ are actuator faults time varying signal ($k < n$), $\bar{D}_i = B_i D_i$, $D_i \epsilon \kappa^{m \times k}$ actuator faults are known.

For the second family, the plant dynamics can be described by

$$\dot{x}(t) = \sum_{i=1}^{p} \mu_i(q(x(t)))[(A_i x_i(t) + B_i u(t) + \bar{D}_i f_a(t)] + \sum_{i=1}^{p} \mu_i(q(x(t)))\Delta A_i x_i(t)$$

$$y(t) = \sum_{i=1}^{p} \mu_i(q(x(t)))[(I + F_s) C_i x(t)] \qquad [5.2]$$

For the third family, the plant dynamics are then given by

$$\dot{x}(t) = \sum_{i=1}^{p} \mu_i(q(x(t)))[(A_i x_i(t) + Bu(t) + \bar{D}_i f_a(t)] + \sum_{i=1}^{p} \mu_i(q(x(t)))\Delta A_i x_i(t)$$

$$y(t) = \sum_{i=1}^{p} \mu_i(q(x(t)))[(I + F_s) C_i x(t)] \qquad [5.3]$$

5.3.1.2. *TS fuzzy observer and the RFFTC scheme design*

In this section, we presented the proposed RFFTC scheme design, FPIEO and FDOS.

5.3.1.2.1. Nonlinear FPIEO and FDOS

This section presents FPIEO and FDOS design methodologies involving actuator faults estimation for TS fuzzy models. PDC structure is employed to achieve the following unknown fuzzy observer structures (Tanaka et al. 1992; Chen et al. 2007). Based on the analysis given in Boukhezzar et al. (2006), Kamal et al. (2011, 2012, 2013), Khan et al. (2011) and Chen et al. (1996), the structure of the FPIEO for estimation of the actuator faults and the dynamic of the fault error estimation *for the first family* can be written as follows:

$$\dot{\hat{x}}_u(t) = \sum_{i=1}^{p} \mu_i [A_i \hat{x}_u(t) + \alpha_i Bu(t) + K_i(y(t) - \hat{y}_u(t)) + \bar{D}_i \hat{f}_a(t)]$$

$$\dot{\hat{f}}_a(t) = \sum_{i=1}^{p} \mu_i L_i (y(t) - \hat{y}_u(t)) = \sum_{i=1}^{p} \mu_i L_i \tilde{y}(t)$$

$$\hat{y}_u(t) = \sum_{i=1}^{p} \mu_i C_i \hat{x}_u(t) \qquad [5.4]$$

where $\hat{x}_u(t)$ is the estimated state vector by unknown fuzzy observer, K_i *(1,2,...,p)* are observation error matrices, L_i are their corresponding integral gains to be determined, $y(t)$ is the output vector and $\hat{y}_u(t)$ is the final output of the unknown fuzzy observer; $\tilde{y}(t) = y(t) - \hat{y}_u(t)$ is the output estimation error and $\hat{f}_a(t)$ is the estimated actuator faults. In the same way and from [5.2] and [5.3], we can induce the inferred output of the FPIEO for the other two cases.

This section also presents FDOS based on the observer stated (Chen et al. 1996; Boukhezzar et al. 2006; Kamal et al. 2011, 2012, 2013; Khan et al. 2011). The inferred modified FDOS states are governed by:

$$\dot{\hat{x}}_o(t) = \sum_{i=1}^{p} \mu_i [A_i \hat{x}_o(t) + \alpha_i Bu(t) + N_i(y(t) - \hat{y}_o(t)) + \overline{D}_i \hat{f}_a(t)]$$

$$\hat{y}_o(t) = \sum_{i=1}^{p} \mu_i C_i \hat{x}_o(t) \qquad [5.5]$$

where $\hat{x}_o(t)$ is the estimated state vector by the FDOS, $\hat{y}_o(t)$ is the final output of the FDOS and $N_i \in \kappa^{n \times g}$ are the FDOS gains. From [5.2] and [5.3], in the same way, we can induce the inferred output of the FDOS for the other two cases.

5.3.1.2.2. The structure of the proposed RFFTC scheme

The proposed scheme is illustrated in Figure 5.1, which detects and isolates the sensor faults and an unknown input observer that estimates the actuator faults and reconstructs the state of the WES from a healthy estimate. Each of the FDOS (observer 1 to observer g linked to fault detection and decision scheme) is driven by all sensors to generate residual signals. The estimated actuator faults from the FPIEO are fed to the FDOS. Through the decision and switcher mechanism, detecting and identifying the faulty sensor is possible (Seron et al. 2008; Jayaram et al. 2010). Finally, by using a switcher, selecting the healthy observer for reconstructing the controller input is enabled.

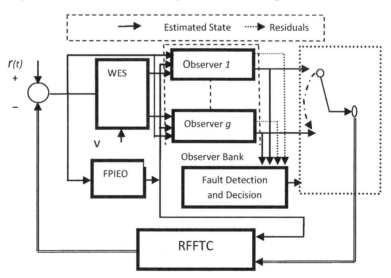

Figure 5.1. *Block diagram of the proposed RFFTC scheme*

5.3.2. *Proposed RFFTC based on FPIEO and FDOS*

In this section, a unique RFFTC synthesis procedure is developed for each member of the TS family to deal with a wide range of uncertainties, sensor faults and actuator faults while maintaining the stability of the closed-loop system.

5.3.2.1. *Nonlinear RFFTC*

For the fuzzy model [5.1], we construct the following RFFTC via the PDC. It is assumed that the fuzzy system [5.1] is locally controllable. A state-feedback with LMIs is used to design a controller for each subsystem.

For the first family, the modified RFFTC is defined by Kamal et al. (2012)

Rule *j*: IF $q_1(x(t))$ is M_{1i} AND ... AND $q_\varepsilon(x(t))$ is $M_{\varepsilon i}$

$$\text{Then } u(t) = -G_j x(t) - D_j \hat{f}_a(t) + r(t)/\alpha_j \quad j = 1,2,...,c \qquad [5.6]$$

From [5.5], the estimated state by the healthy observer from FDOS is used, then the final output of the modified RFFTC becomes

$$u(t) = \frac{\sum_{j=1}^{c} \mu_j [-G_j \hat{x}_o(t) - D_j \hat{f}_a(t) + r(t)]}{\sum_{j=1}^{c} \alpha_j \mu_j} \qquad [5.7]$$

For the second family, the overall output of the RFFTC is given by

$$\dot{u}(t) = z_k u(t) + \sum_{j=1}^{c} \mu_j [-G_j \hat{x}_o(t) - D_j \hat{f}_a(t) + r(t)] \qquad [5.8]$$

For the last family, the RFFTC is given by

$$u(t) = \sum_{j=1}^{c} \mu_j [-G_j x_o(t) - D_j \hat{f}_a(t) + r(t)] \qquad [5.9]$$

5.3.2.2. *The augmented fuzzy control system*

In order to carry out the analysis, the closed-loop fuzzy system should be obtained first by establishing the conditions for the asymptotic convergence of the observers [5.4] and [5.5].

For the first family, the fuzzy control system of the state and the errors can be obtained.

Let $e_1(t) = x(t) - \hat{x}_o(t)$ [5.10]

$$\dot{x}(t) = \sum_{i=1}^{p} \mu_i A_i x_i(t) + \left\{ \sum_{i=1}^{p} \alpha_i \mu_i \right\} Bu(t) + \sum_{i=1}^{p} \mu_i \bar{D}_i f_a(t) + \sum_{i=1}^{p} \mu_i \Delta A_i x_i(t) \quad [5.11]$$

With the modified TS fuzzy FTC [5.7] employed, the TS fuzzy system [5.10] has the following closed-loop:

$$\dot{x}(t) = \sum_{i=1}^{p} \mu_i A_i x(t) + \sum_{i=1}^{p} \mu_i \bar{D}_i f_a(t) + \sum_{i=1}^{p} \mu_i \Delta A_i x_i(t)$$

$$+ \left\{ \sum_{i=1}^{p} \alpha_i \mu_i \right\} B \left\{ \sum_{j=1}^{p} \mu_j [-G_j \hat{x}_o(t) - D_j \hat{f}_a(t) + r(t)] \right\} / \left\{ \sum_{j=1}^{p} \alpha_j \mu_j \right\} \quad [5.12]$$

Let $\tilde{f}_a(t) = f_a(t) - \hat{f}_a(t)$ [5.13]

From [5.10] and [5.13], a TS fuzzy closed-loop can be observed:

$\dot{x}(t) =$

$$\sum_{i=1}^{p} \sum_{j=1}^{c} \mu_i \mu_j [(A_i - BG_j) x_i(t) + BG_j e_1(t) + \bar{D}_j \tilde{f}_a(t) + Br(t)] + \sum_{i=1}^{p} \mu_i \Delta A_i x_i(t) \quad [5.14]$$

By taking the derivative of [5.10] and substituting from [5.1], [5.6] and [5.13], the following is obtained:

$$\dot{e}_1(t) = \sum_{i=1}^{p} \sum_{j=1}^{c} \mu_i \mu_j [(\Delta A_i - N_i F_s C_j) x_i(t) + (A_i - N_i C_j) e_1(t) + \bar{D}_i \tilde{f}_a(t)] \quad [5.15]$$

Using $e_2(t) = x(t) - \hat{x}_u(t)$ [5.16]

then taking the derivative of [5.16] and substituting from [5.16], [5.4] and [5.13], the following is obtained:

$$\dot{e}_2(t) = \sum_{i=1}^{p}\sum_{j=1}^{c} \mu_i \mu_j [(\Delta A_i - K_i F_s C_j) x_i(t) + (A_i - K_i C_j) e_2(t) + \bar{D}_i \tilde{f}_a(t)] \quad [5.17]$$

we assume the actuator fault is not constant but time varying, the derivative of $\tilde{d}(t)$ can be written as

$$\dot{\tilde{f}}_a(t) = \dot{f}_a(t) - \dot{\hat{f}}_a(t) = \dot{f}_a(t) - \sum_{i=1}^{p} \mu_i [L_i C_i e_2(t) + L_i F_s C_i x(t)] \quad [5.18]$$

Equations [5.14], [5.15], [5.17] and [5.18] can be written as

$$\dot{X}(t) = \sum_{i=1}^{p}\sum_{j=1}^{c} \mu_i \mu_j (H_{ij} + \Delta H_{ij}) X(t) + Sr(t) + \Psi_a \varphi(t) \quad [5.19]$$

with $X(t) = \begin{bmatrix} x(t) \\ e_1(t) \\ e_2(t) \\ \tilde{f}_a(t) \end{bmatrix}, \varphi(t) = \begin{bmatrix} \dot{f}_a(t) \end{bmatrix}, S = \begin{bmatrix} B \\ 0 \\ 0 \\ 0 \end{bmatrix}, \Psi_a = \begin{bmatrix} 0 \\ 0 \\ 0 \\ I \end{bmatrix},$

$$H_{ij} = \begin{bmatrix} (A_i - BG_j) & BG_j & 0 & \bar{D}_j \\ 0 & (A_i - N_i C_j) & 0 & \bar{D}_i \\ 0 & 0 & (A_i - K_i C_j) & \bar{D}_i \\ 0 & 0 & -L_i C_i & 0 \end{bmatrix},$$

$$\Delta H_{ij} = \begin{bmatrix} \Delta A_i & 0 & 0 & 0 \\ \Delta A_i - N_i F_s C_j & 0 & 0 & 0 \\ \Delta A_i - K_i F_s C_j & 0 & 0 & 0 \\ L_i F_s C_i & 0 & 0 & 0 \end{bmatrix}$$

For the second family, when the input matrices are different, using [5.2], [5.8], [5.15], [5.17] and [5.18], the augmented fuzzy system is given by

$$\dot{X}(t) = \sum_{i=1}^{p} \sum_{j=1}^{c} \mu_i \mu_j (H_{ij} + \Delta H_{ij})X(t) + S_i r(t) + \Psi_a \varphi(t) \quad [5.20]$$

with $X(t) = \begin{bmatrix} x(t) \\ u(t) \\ e_1(t) \\ e_2(t) \\ \tilde{f}_a(t) \end{bmatrix}$, $\varphi(t) = \begin{bmatrix} \dot{f}_a(t) \\ \hat{f}_a(t) \end{bmatrix}$, $S_i = \begin{bmatrix} 0 \\ B_i \\ 0 \\ 0 \\ 0 \end{bmatrix}$, $\Psi_a = \begin{bmatrix} 0 & \bar{D}_i \\ 0 & 0 \\ 0 & 0 \\ 0 & 0 \\ I & 0 \end{bmatrix}$,

$$H_{ij} = \begin{bmatrix} A_i & B_i & 0 & 0 & \bar{D}_j \\ G_j & Z_k & -G_j & 0 & \bar{D}_j \\ 0 & 0 & (A_i - N_i C_j) & 0 & \bar{D}_i \\ 0 & 0 & 0 & (A_i - K_i C_j) & \bar{D}_i \\ 0 & 0 & 0 & -L_i C_j & 0 \end{bmatrix}$$

$$\Delta H_{ij} = \begin{bmatrix} \Delta A_i & 0 & 0 & 0 & 0 \\ 0 & 0 & 0 & 0 & 0 \\ \Delta A_i - N_i F_s C_j & 0 & 0 & 0 & 0 \\ \Delta A_i - K_i F_s C_j & 0 & 0 & 0 & 0 \\ L_i F_s C_i & 0 & 0 & 0 & 0 \end{bmatrix}$$

In the same way, the augmented fuzzy system for the third family can be deduced.

5.3.3. *Proposed RFFTC stability and robustness analysis*

The analysis procedures are also the same as those given in Boukhezzar et al. (2006), Kamal et al. (2011, 2012, 2013), Khan et al. (2011) and Chen et al. (1996), and so the analysis results will be presented without proof. The main result for the global asymptotic stability of a TS fuzzy model with parameter uncertainties, sensor faults and actuator faults are summarized by the following theorem.

THEOREM 5.1.– *The fuzzy control system as given by [5.19] is stable if the controller and the observer gains are set to $G_j = M_{a11}^{-1} Y_j$ and $N_i = P_{a22}^{-1} O_i$ and $\bar{E}_i = P_2^{-1} X_i$ with the matrices X_i, M_{a11}, Y_j and O_i satisfying the following LMIs:*

$$M_{a11} A_i^T + A_i M_{a11} - (B_i Y_j)^T - (B_i Y_j) < 0 \qquad [5.21]$$

$$A_i^T P_{a22} + P_{a22} A_i - (O_i C_j)^T - (O_i C_j) < 0 \qquad [5.22]$$

$$H_{bij}^T P_2 + P_2 H_{bij} - (X_i \bar{C}_j)^T - (X_i \bar{C}_j) < 0 \qquad [5.23]$$

where $M_{a11} = P_{a11}$, $H_{bij} = \begin{bmatrix} A_i & \bar{D}_i \\ 0 & 0 \end{bmatrix}$, $\bar{E}_i = \begin{bmatrix} K_i \\ L_i \end{bmatrix}$, $\bar{C}_j = \begin{bmatrix} C_j \\ 0 \end{bmatrix}^T$,

$$P = \begin{bmatrix} P_1 & 0 \\ 0_{2x2} & P_2^{2x2} \end{bmatrix}, \quad P_1 = \begin{bmatrix} P_{a11} & 0 \\ 0 & P_{a22} \end{bmatrix}, \qquad [5.24]$$

5.3.4. *WES with DFIG application*

The WES model system with DFIG system used in Boukhezzar et al. (2006), Kamal et al. (2011, 2012, 2013), Khan et al. (2011) and Chen et al. (1996) will be used as an application example. The nonlinear systems are subject to large parameter uncertainties, actuator faults and sensor faults. The objective here is to conceive an actuator and sensor fault-tolerant control for WES with parameters uncertainties within 35% of the nominal values and disturbance (20% of wind speed). Based on Boukhezzar et al. (2006), Kamal et al. (2011, 2012, 2013), Khan et al. (2011) and Chen et al. (1996), we can generalize that the ith rule of the continuous TS fuzzy models including actuator faults and sensor faults are of the following forms:

Rule i: IF $q_1(x_5(t))$ is $M_{\ni i}$ and $q_2(x_6(t))$ is $D_{\ni i}$ and $q_3(x_3(t))$ is $N_{\ni i}$

then $\dot{x}(t) = (A_i + \Delta A_i)x(t) + B_i u(t) + \bar{D}_i f_a(t)$

$$y(t) = (I + F_s)C_i x(t) \quad i = 1,2,\ldots,8 \; ; \; \jmath = 1,2 \qquad [5.25]$$

where the subsystems are determined as: $\bar{D}_i = \begin{bmatrix} 0 & 0 & 0 & 1 & 0 \\ 0 & 1 & 0 & 1 & 0 \end{bmatrix}^T$

$$A_i = \begin{bmatrix} \dfrac{-R_s}{\xi L_s} & \omega_s + \dfrac{(1-\xi)n_p q_{2i}}{\xi} & \dfrac{L_m R_r}{\xi L_r L_s} & \dfrac{L_m}{\xi L_s} n_p q_{2i} & 0 & 0 & 0 & 0 \\[6pt] \dfrac{-R_s}{\xi L_s} & -(\omega_s + \dfrac{1-\xi}{\xi}n_p q_{2i}) & \dfrac{R_r L_m}{\xi L_r L_s} & -\dfrac{L_m}{\xi L_s} n_p q_{2i} & 0 & 0 & 0 & 0 \\[6pt] \dfrac{R_s L_m}{\xi L_r L_s} & -\dfrac{L_m}{\xi L_r} n_p q_{2i} & -\dfrac{R_r}{\xi L_r} & \omega_s - \dfrac{1}{\xi}n_p q_{2i} & 0 & 0 & 0 & 0 \\[6pt] \dfrac{R_s L_m}{\xi L_r L_s} & \dfrac{L_m}{\xi L_r} n_p q_{2i} & -\dfrac{R_r}{\xi L_r} & -\omega_s + \dfrac{1}{\xi}n_p q_{2i} & 0 & 0 & 0 & 0 \\[6pt] 0 & 0 & 0 & 0 & (\dfrac{D_r}{J_r} + \dfrac{K_{opt}}{J_r} q_{1i}) & 0 & -\dfrac{n_b}{J_r} & 0 \\[6pt] 0 & 0 & 0 & 0 & 0 & \dfrac{-D_g}{J_g} & \dfrac{1}{J_g} & -\dfrac{1}{J_g} \\[6pt] 0 & 0 & 0 & 0 & a75 + \dfrac{D_{lse} K_{opt}}{n_b J_r} q_{1i} & a76 & a77 & \dfrac{D_{ls}}{n_b^2 J_g} \\[6pt] 0 & 0 & 0 & 0 & 0 & 0 & 0 & -\dfrac{1}{\tau_g} \end{bmatrix},$$

$$B_i = \begin{bmatrix} -\dfrac{1}{\xi L_s} & 0 & \dfrac{L_m}{\xi L_r L_s} & 0 & 0 \\ 0 & -\dfrac{1}{\xi L_s} & 0 & \dfrac{L_m}{\xi L_r L_s} & 0 \\ \dfrac{L_m}{\xi L_r L_s} & 0 & -\dfrac{1}{\xi L_r} & 0 & 0 \\ 0 & \dfrac{L_m}{\xi L_r L_s} & 0 & -\dfrac{1}{\xi L_r} & 0 \\ 0 & 0 & 0 & 0 & 0 \\ 0 & 0 & 0 & 0 & 0 \\ 0 & 0 & 0 & 0 & 0 \\ 0 & 0 & 0 & 0 & \dfrac{1}{\tau_g} \end{bmatrix},$$

$$C_i = \begin{bmatrix} 0 & 0 & 0 & 0 & 1 & 0 & 0 \\ 0 & 0 & 0 & 0 & 0 & 1 & 0 \\ 0 & 0 & 0 & -\dfrac{L_m V_s}{L_s} & 0 & 0 & 0 \\ 0 & 0 & \dfrac{V_s^2}{\omega_s L_s} q_{3i} & -\dfrac{L_m V_s}{L_s} & 0 & 0 & 0 & 0 \end{bmatrix}$$

where ΔA_i ($i = 1,2,...,8$) represent the non-time varying system parameters uncertainties (J_r, J_g, R_s and R_r) but bounded; the elements of ΔA_i ($i = 1,2,...,8$) randomly achieve the values within 35% of their nominal values corresponding to A_i ($i = 1,2,...,8$); ΔB_i ($i = 1,2,...,8$) = 0. The actuator faults $f_a(t) = [f_{a1}(t)\ f_{a2}(t)]^T$ are time varying and defined as follows:

$$f_{a1}(t) = \begin{cases} 0 & t<120\,\text{s} \\ 4\sin(\pi t) & t \geq 120\,\text{s} \end{cases},\ f_{a2}(t) = \begin{cases} 0 & t<120\text{ s} \\ 2\sin(\pi t) & t \geq 120\text{ s} \end{cases} \quad [5.26]$$

Faults are modeled as additive signals to generator speed sensor outputs based on studies of Boukhezzar et al. (2006), Kamal et al. (2011, 2012, 2013), Khan et al. (2011) and Chen et al. (1996).

According to the analysis above, the procedure for finding the proposed RFFTC with a nonlinear unknown input observer and the FDOS observer is summarized as follows:

1) obtain the mathematical model of the WES to be controlled;

2) obtain the fuzzy plant model for the system stated in step (1) by means of a fuzzy modeling method;

3) solve LMIs [5.21]–[5.23] to obtain X_i, Y_j, P_{a22}, P_2, K_i, L_i, N_i and $\overline{E}_i$, thus $G_j = M_{a11}^{-1} Y_j$ and $N_i = P_{a22}^{-1} O_i$ and $\overline{E}_i = P_2^{-1} X_i$;

4) construct fuzzy observers [5.4], [5.5] and the fuzzy controller [5.7]–[5.9].

5.3.5. Simulations and results

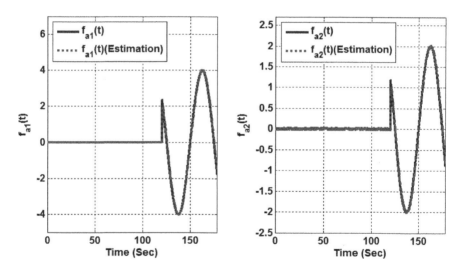

Figure 5.2. Actuator faults ($f_{a1}(t)$ and $f_{a2}(t)$) and their estimations

The proposed controller for the WES is tested for the random variation of wind speed to illustrate the effectiveness of the proposed method. The control objective of this chapter is to design a RFFTC law to ensure that all signals in the closed-loop system are bounded. Figure 5.2 shows the time evolution of the time varying actuator faults $f_a(t)$ and its estimate $\hat{f}_a(t)$,which occur after 120 seconds (s). Figure 5.3 shows the proportional signals that represent sensor failures, which has been added to the output of sensor 6 between $t = 40$ s and $t = 80$ s. The magnitudes of faults are between ≈10% to ≈20% of the nominal values of the variables and the parametric uncertainties J_r, J_g, R_s and R_r are considered within 35% of their nominal values.

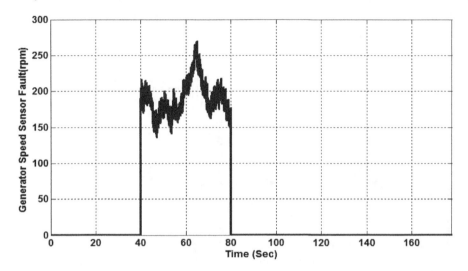

Figure 5.3. *Proportional error on the generator speed sensor*

The simulation results are given in Figures 5.4–5.7 with (left) and without (right) the RFFTC strategy and all the simulations are realized on the nonlinear model given in Boukhezzar et al. (2006), Kamal et al. (2011, 2012, 2013), Khan et al. (2011) and Chen et al. (1996), with the presence of non-time varying parametric uncertainties, sensor faults and time varying actuator faults. In Figures 5.4–5.7 (left), the control law is based on one observer (observer 6) without using the switching block and the unknown observers. We can observe that between $t = 0$ s to $t ≈ <40$ s and $t ≈ >80$ sec to $t ≈ <120$ s, there are no faults, providing good tracking performance. However, between $t = 40$ s to $t = 80$ s and after $t = 120$ s, we can see that the WES's performance is reduced right after the generator speed sensor and actuators have become faulty. Figures 5.4 and 5.5 (right) shows WES state variables and their estimated signals, when the law control is based on the bank observers (observer 5) with the switch block and the unknown observers. We can note that the WES remains stable despite the presence of faults and parameter uncertainties, which shows the effectiveness of the proposed RFFTC strategy. The rotational speed of the WT and generator (dashed line), respectively, and their estimates (dotted lines) in the presence of parametric uncertainties, sensor faults and actuator faults are shown in Figures 5.5 and 5.5. In order to obtain optimality, the $r(t) = \Omega_{gref} = \Omega_{opt} = \omega_s - n_b\lambda_{opt}V/R$ profiles are chosen in such a way to follow the optimal tip speed ratio (λ_{opt}) (solid line).

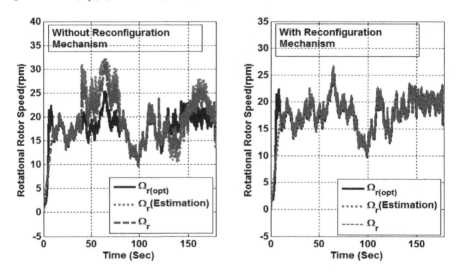

Figure 5.4. *The trajectories of Ω_r and its estimate without (left) and with (right) RFFTC strategy*

The switching from observer 6 to observer 5 is visualized clearly at $t \approx 40$ s (Figures 5.4–5.7 (right)). We note that switching observers is carried out without loss of control of the system's state.

Figures 5.6 and 5.7 show the active power P_s (dashed line) and its estimate (dotted line) and the reactive power Q_s (dashed line) and its estimate (dotted line), respectively. It can be seen that our design can have good tracking performance, although there are severe actuator faults and sensor fault entering into the system.

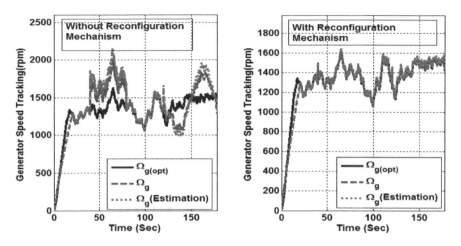

Figure 5.5. *The trajectories of Ω_g and its estimate without (left) and with (right) RFFTC strategy*

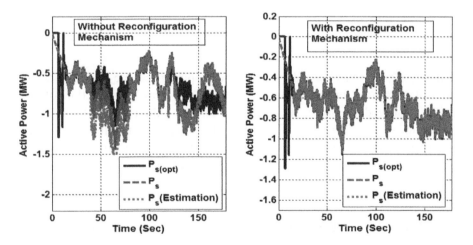

Figure 5.6. *The trajectories of P_s and its estimate without (left) and with (right) RFFTC strategy*

From the simulation, it can be seen that without the reconfiguration mechanism, the WES lost its performance after the generator speed sensor and the actuator became faulty. Although for the same reference input and by using the RFFTC scheme strategy proposed, the WES remains stable in the presence of sensor faults, parameter uncertainties and actuator faults. This demonstrates the effectiveness of the proposed RFFTC strategy.

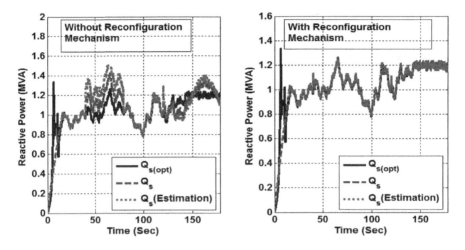

Figure 5.7. *The trajectories of Q_s (dashed line) and its estimate (dotted line) without (left) and with (right) RFFTC strategy*

In summary, it has been shown that the proposed scheme is able to detect and isolate sensor faults, through a proper and feasible selection of the healthy observed variables. It can also compensate the actuator faults using the nonlinear unknown input observers. The simulation results demonstrate the effectiveness of the proposed control approach. The proposed control scheme can guarantee the stability of the closed-loop system and the convergence of the output tracking error.

5.4. RFSFTC of WES with DFIG subject to sensor and actuator faults

This chapter proposes an FDOS method using a nonlinear FPIEO with a RFSFTC algorithm for fuzzy TS systems subject to sensor faults, parametric uncertainties and time varying actuator faults. FDOS provide residuals for detection and isolation of sensor faults, which can affect a TS model. The TS fuzzy model is adopted for fuzzy modeling of the uncertain nonlinear system and establishing fuzzy state observers. The concept of PDC is employed to design RFSFTC and fuzzy observers from the TS fuzzy models. TS fuzzy systems are classified into three families based on the input matrices and an

RFSFTC synthesis procedure is given for each family. In each family, sufficient conditions are derived for robust stabilization, in the sense of Taylor series stability and the Lyapunov method, for the TS fuzzy system with parametric uncertainties, sensor faults and actuator faults. The sufficient conditions are formulated in the format of linear matrix equalities (LMEs). The effectiveness of the proposed controller design methodology is finally demonstrated through a WES with doubly DFIG to illustrate the effectiveness of the proposed method.

This section is organized as follows. Section 5.4.1 provides the proposed RFSFTC scheme, TS fuzzy model FDOS and FPIEO. Section 5.4.2 shows the stability and robustness conditions for the proposed algorithm followed by the calculation of state RFSFTC, FDOS and FPIEO gains. Section 5.4.3 shows simulation and results on WES model system with DFIG.

5.4.1. *TS fuzzy plant model with actuator faults, sensor faults and parameter uncertainties*

In this section, the TS fuzzy plant model subject to parameter uncertainties, sensor faults and actuator faults will be expressed as a weighted sum of a number of fuzzy systems. An augmented TS fuzzy plant model is formed by adding a fuzzy uncertainty regenerator. The inferred outputs of the fuzzy scheduler fault system for three families have been introduced in section 5.3. Since this chapter considers the occurrence of sensor faults and actuator faults in the presence the parameter uncertainties, for the first family based on the study of Boukhezzar et al. (2006), Kamal et al. (2011 2012, 2013), Khan et al. (2011) and Chen et al. (1996), the dynamics of fuzzy scheduler fault system for the first family are given by

$$\dot{x}(t) = \sum_{i=1}^{p} \sum_{l=1}^{s} \mu_i h_l [(A_i + \Delta \tilde{A}_l) x(t) + \alpha_i B u(t) + \bar{D}_i f_a(t)]$$

$$y(t) = \sum_{i=1}^{p} \mu_i(q(t))(I + F_s) C_i x(t) \qquad [5.27]$$

In the same way, we can induce the inferred output of the fuzzy scheduler for the other two cases.

5.4.2. Proposed RFSFTC algorithm based on FPIEO and FDOS

In this section, RFSFTC is developed for each member of the TS family to deal with a wide range of parameter uncertainties, sensor faults and actuator faults such that the closed-loop system is stable. We will use the same proposed scheme, as illustrated in Figure. 5.1, but we will change the RFFTC to RFSFTC.

5.4.2.1. Proposed RFSFTC Controllers

Based on the analysis in Boukhezzar et al. (2006), Kamal et al. (2011, 2012, 2013), Khan et al. (2011) and Chen et al. (1996) in obtaining the RFSFTC and based on the FPIEO [5.4] and FDOS [5.5], we can obtain the fuzzy scheduler FTC. *For the first family*, the inferred output of the modified RFSFTC in Boukhezzar et al. (2006), Kamal et al. (2011, 2012, 2013), Khan et al. (2011) and Chen et al. (1996) will be as follows:

$$u(t) = \frac{\sum_{j=1}^{c}\sum_{l=1}^{s} \mu_j h_l [-G_{jl}\hat{x}_o(t) - D_j \hat{f}_a(t) + r(t)]}{\sum_{l=1}^{s}\sum_{j=1}^{c} \alpha_j \mu_j h_l} \qquad [5.28]$$

In the same way, we can induce the inferred output of the RFSFTC for the other two cases as follows:

– For the second family, the modified RFSFTC in some previous studies (Boukhezzar et al. 2006; Kamal et al., 2011, 2012, 2013; Khan 2011; Chen et al. 1996) becomes

$$\dot{u}(t) = Z_k u(t) + \sum_{j=1}^{c}\sum_{l=1}^{s} \mu_j h_l [-G_{jl}\hat{x}_o(t) - D_j \hat{f}_a(t) + r(t)] \qquad [5.29]$$

– For the last family, the RFSFTC in Boukhezzar et al. (2006), Kamal et al. 2011, 2012, 2013), Khan et al. (2011) and Chen et al. (1996) becomes

$$u(t) = \sum_{j=1}^{c}\sum_{l=1}^{s} \mu_j h_l [-G_{jl}\hat{x}_o(t) - D_j \hat{f}_a(t) + r(t)] \qquad [5.30]$$

5.4.2.2. Stability analysis for the proposed RFSFTC algorithm

In order to carry out the analysis, the closed-loop fuzzy system should be obtained first by establishing the conditions for the asymptotic convergence of the observers.

For the first family, based on [5.10], [5.13], [5.16] and [5.27], we can obtain the augmented fuzzy system.

$$\dot{X}(t) = \sum_{i=1}^{p}\sum_{j=1}^{p}\sum_{l=1}^{s} \mu_i \mu_j h_l [(H_{ijl} + \Delta \tilde{H}_{ijl})X(t) + Sr(t) + \Psi_a \varphi(t)] \qquad [5.31]$$

with $X(t) = \begin{bmatrix} x(t) \\ e_1(t) \\ e_2(t) \\ \tilde{f}_a(t) \end{bmatrix}$, $\varphi(t) = \begin{bmatrix} \dot{f}_a(t) \end{bmatrix}$, $\Delta H_{ijl} = \begin{bmatrix} \Delta \tilde{A}_l & 0 & 0 & 0 \\ \Delta \tilde{A}_l - N_i F_s C_j & 0 & 0 & 0 \\ \Delta \tilde{A}_l - K_i F_s C_j & 0 & 0 & 0 \\ L_i F_s C_j & 0 & 0 & 0 \end{bmatrix}$, $S = \begin{bmatrix} B \\ 0 \\ 0 \\ 0 \end{bmatrix}$,

$\Psi_a = \begin{bmatrix} 0 \\ 0 \\ 0 \\ I \end{bmatrix}$, $H_{ijl} = \begin{bmatrix} (A_i - BG_{jl}) & BG_{jl} & 0 & \bar{D}_j \\ 0 & (A_i - N_i G_{jl}) & 0 & \bar{D}_i \\ 0 & 0 & (A_i - K_i C_j) & \bar{D}_i \\ 0 & 0 & -L_i C_i & 0 \end{bmatrix}$

In the same way, the augmented fuzzy system for the second and the third families can be deduced.

5.4.3. Derivation of the stability and robustness conditions

Based on the analysis before, the main result for the global asymptotic stability of a TS fuzzy model with parameter uncertainties, sensor faults and actuator faults input are summarized by the following theorem.

THEOREM 5.2.– *The fuzzy control system as given by [5.31] is stable if the controller and the observer gains are set to* $G_j = M_{a11}^{-1} Y_j$ *and* $N_i = P_{a22}^{-1} O_i$

and $\bar{E}_i = P_2^{-1} X_i$ with the matrices X_i, M_{a11}, Y_j and O_i satisfying the following LMEs:

$$M_{a11} A_i^T + A_i M_{a11} - (B_i Y_{jl})^T - (B_i Y_{jl}) = -\sigma I \qquad [5.32]$$

$$A_i^T P_{a22} + P_{a22} A_i - (O_i C_j)^T - (O_i C_j) = -\sigma I \qquad [5.33]$$

$$H_{bi}^T P_2 + P_2 H_{bi} - (X_i \bar{C}_j)^T - (X_i \bar{C}_j) = -\sigma I \qquad [5.34]$$

where $P = \begin{bmatrix} P_1 & 0_{2 \times 2} \\ 0_{2 \times 2} & P_2 \end{bmatrix}$, $H_{bi} = \begin{bmatrix} A_i & \bar{D}_i \\ 0 & 0 \end{bmatrix}$, $\bar{E}_i = \begin{bmatrix} K_i \\ L_i \end{bmatrix}$, $\bar{C}_j = \begin{bmatrix} C_j \\ 0 \end{bmatrix}^T$,

$P_1 = \begin{bmatrix} P_{a11} & 0 \\ 0 & P_{a22} \end{bmatrix}$.

$M_{a11} = P_{a11}^{-1}$, σ is the robustness index and is determined by the same manner in Boukhezzar et al. (2006), Kamal et al. (2011, 2012, 2013), Khan et al. (2011) and Chen et al. (1996). According to the analysis above, the procedure for finding the RFSFTC and the fuzzy observer are summarized as follows:

– obtain the mathematical model of the uncertain multivariable nonlinear plant;

– obtain the TS fuzzy plant model (with parameter uncertainty information) of the plant obtained in the previous step;

– determine the ranges of the parameter uncertainties, that is, ΔA, so as to define the specified parameter space;

– model the parameter uncertainties with a fuzzy uncertainty;

– build the fuzzy scheduler using the same fuzzy rule antecedents of the fuzzy uncertainty regenerator;

– solve [5.32]–[5.34] to obtain X_i, Y_j, P_{a22}, P_2, K_i, L_i, N_i and $\bar{E}_i$ thus ($G_{jl} = Y_{jl} M_{a11}^{-1}$ and $N_i = P_{a22}^{-1} O_i$ and $\bar{E}_i = P_2^{-1} X_i$);

– construct the fuzzy observers [5.4], [5.5] and the RFSFTC [5.28]–[5.30].

5.4.4. WES with DFIG application and simulations and results

We use the random variation of wind speed as shown in Boukhezzar et al. (2006), Kamal et al. (2011, 2012, 2013), Khan et al. (2011) and Chen et al. (1996). It is used to test the proposed controller and the WES model system with DFIG system. The nonlinear systems are subject to actuator and sensor faults. The actuator fault $f_a(t)$, as shown in Figure 6.8(left), represents the actuator fault signals. For the testing purpose, it is required that at least one sensor fails every time. Faults are modeled as proportional signals to sensor outputs as shown in Figure 5.8 (right) and the parametric uncertainties R_s and R_r are considered within 40% of their nominal values. The actuator fault $f_a(t)$ is time varying and defined as follows:

$$f_a(t) = \begin{cases} 0 & t<40s \\ 4\sin(\pi t) & t \geq 40s \end{cases} \quad [5.35]$$

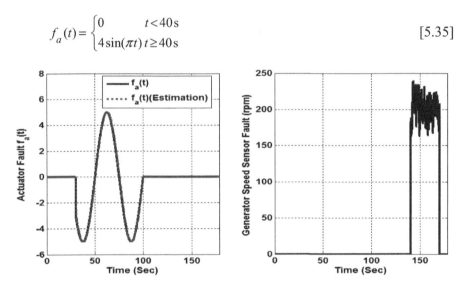

Figure 5.8. *The actuator fault $f_a(t)$ and its estimate (left) and the proportional error on the generator speed sensor (right)*

Figures 5.9 and 5.10 show the rotational speed of the WT and generator (dashed line), respectively, and their estimates (dotted lines) in the presence of parametric uncertainties, sensor faults and actuator faults. A simulation without the RFSFTC scheme strategy is shown on the left and one with the RFSFTC scheme strategy is shown on the right. In order to obtain optimality, the $r(t) = \Omega_{gref} = \Omega_{opt} = \omega_s\text{-}n_b\lambda_{opt}v/R$ profile is chosen in such a way as to follow the optimal tip speed ratio (λ_{opt}) (solid line). From the simulation results using the proposed control scheme, it can be seen that the outputs of the system are bounded and good tracking performance can be obtained through the uncertain nonlinearities of the system, sensor faults and the actuator fault.

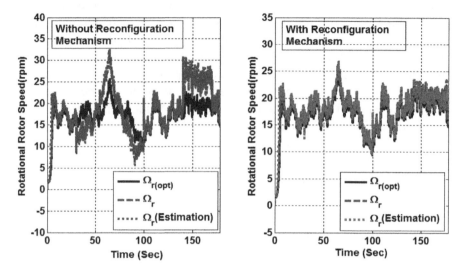

Figure 5.9. *The trajectories of Ω_r and its estimate without (left) and with (right) RFSFTC strategy*

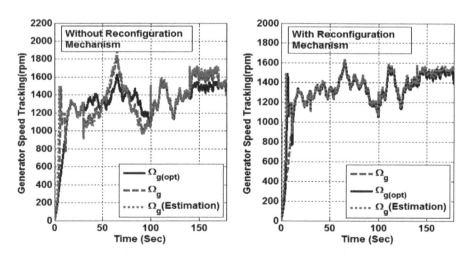

Figure 5.10. *The trajectories of Ω_g and its estimate without (left) and with (right) RFSFTC strategy*

Figures 5.11 and 5.12 show the active power P_s (dashed line) and its estimate (dotted line) and the reactive power Q_s (dashed line) and its estimate (dotted line), respectively, without RFSFTC scheme strategy (left) and with RFSFTC scheme strategy (right).

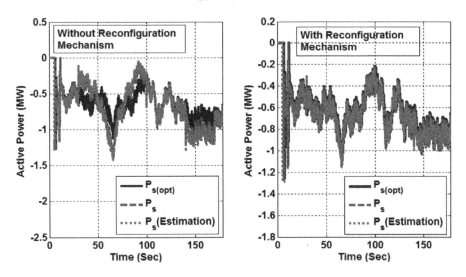

Figure 5.11. *The trajectories of P_s and its estimate without (left) and with (right) RFSFTC strategy*

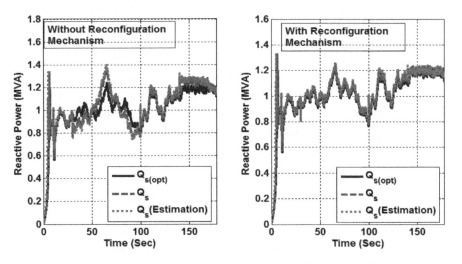

Figure 5.12. *The trajectories of Q_s (dashed line) and its estimate (dotted line) without (left) and with (right) RFSFTC strategy*

From the simulation, it can be seen that without the reconfiguration mechanism, the WES lost performance just after the generator speed sensor became faulty, whereas for the same reference input and by using the RFSFTC scheme strategy proposed, the WES remains stable in the presence of sensor faults, parameter uncertainties and actuator fault that demonstrates the effectiveness of the proposed RFSFTC strategy.

In summary, it has been shown that the proposed scheme is able to detect and isolate sensor faults through a proper and feasible selection of the healthy observed variables. It can also compensate for the actuator fault using the nonlinear FPIEO. The simulation results demonstrate the effectiveness of the proposed control approach. The proposed control scheme can guarantee the stability of the closed-loop system and the convergence of the output tracking error.

5.5. RDFFTC of hybrid wind-diesel storage system subject to actuator and sensor faults

This section extends the general ideas proposed in Boukhezzar et al. (2006), Kamal et al. (2011, 2012, 2013), Khan et al. (2011) and Chen et al. (1996). A fuzzy FTC approach is proposed for WES with norm-bounded parameter uncertainties, sensor faults and actuator faults. The algorithm utilizes fuzzy systems based on TS fuzzy models to approximate uncertain nonlinear systems with sensor faults and actuator faults. Sufficient stabilization conditions of the fuzzy fault-tolerant control systems are given, which are formulated in terms of LMEs. The proposed algorithm combines the merits of (i) the states of the closed-loop system that will follow those of a user-defined stable reference model despite the presence of bounded magnitude sensor faults, actuator faults and parameter uncertainties, and (ii) the algorithm that maximizes the power coefficient for a fixed pitch and reduces the voltage ripple.

This section is organized as follows. Section 5.5.1 describes the fuzzy system subject to sensor and actuator faults with parameter uncertainties. The proposed FTC and the condition for stability are presented in section 5.5.2. Section 5.5.3 presents TS fuzzy description for the WES and simulations of sensor and actuator faults and results analysis.

5.5.1. *Fuzzy observer scheme for the uncertain system with sensor and actuator faults*

The objective is to design a fuzzy observer for a nonlinear system with parametric uncertainties, sensor faults and actuator faults. Considering initially the ith rule of the TS fuzzy model with parametric uncertainties, sensor and actuator faults (Chen et al. 1996; Boukhezzar et al. 2006; Kamal et al. 2011, 2012, 2013; Khan et al. 2011) are given by:

Plant rule i: IF $q_1(x(t))$ is M_{1i} AND ... AND $q_\varepsilon(x(t))$ is $M_{\varepsilon i}$

Then $\dot{x}(t) = (A_{fi} + \Delta A_{fi})x(t) + B_{fi}U(t) + E_{ai}f_a(t)$,

$$y(t) = C_{fi}x(t) + E_{si}f_s(t) \quad i = 1,\ldots,p \qquad [5.36]$$

where $q(x(t)) = [q_1(x(t)),\ldots,q_\varepsilon(x(t))]$ are also measurable variables and do not depend on the faults (Kamal et al. 2012), that is, the premise variables, $x(t)\in\kappa^{nx1}$ is the state vector, $U(t)\in\kappa^{mx1}$ is the control input vector, $y(t)\in\kappa^{gx1}$ is the output vector, $A_{fi}\in\kappa^{nxn}$, $B_{fi}\in\kappa^{nxm}$ and $C_{fi}\in\kappa^{gxn}$ are system input and output matrices, respectively, $\Delta A_{fi}\in\kappa^{nxn}$ is the parametric uncertainties in the plant model, E_{ai} and E_{si} are actuator and sensor faults matrices, respectively, and $f_a(t)$ and $f_s(t)$ represents actuator and sensor faults, respectively, which is assumed to be bounded. The inferred system is given by

$$\dot{x}(t) = \sum_{i=1}^{p} \mu_i[(A_{fi} + \Delta A_{fi})x(t) + B_{fi}U(t) + E_{ai}f_a(t)],$$

$$y(t) = \sum_{i=1}^{p} \mu_i[C_{fi}x(t) + E_{si}f_s(t)] \qquad [5.37]$$

Consider also the state $Z(t)\in\kappa^{rx1}$ that is a filtered version of the output $y(t)$ [5.12]. This state is given by:

$$\dot{Z}(t) = \sum_{i=1}^{p} \mu_i[-A_{zi}z(t) + A_{zi}C_{fi}x(t) + A_{zi}E_{si}f_s(t)] \qquad [5.38]$$

where $-A_{zi}\mathcal{C}\kappa^{rxr}$ is the stable matrix; from the [5.37] and [5.38], we can obtain the augmented system:

$$\dot{X}(t) = \sum_{i=1}^{p} \mu_i[(A_i + \Delta A_i)X(t) + B_i U(t) + E_i f(t)],$$

$$Y(t) = \sum_{i=1}^{p} \mu_i C_i X(t) \qquad [5.39]$$

where $X(t) = \begin{bmatrix} x(t) \\ Z(t) \end{bmatrix}$, $f(t) = \begin{bmatrix} f_a(t) \\ f_s(t) \end{bmatrix}$, $A_i = \begin{bmatrix} A_{fi} & 0 \\ A_{zi}C_{fi} & -A_{zi} \end{bmatrix}$, $\Delta A_i = \begin{bmatrix} \Delta A_{fi} & 0 \\ 0 & 0 \end{bmatrix}$,

$B_i = \begin{bmatrix} B_{fi} \\ 0 \end{bmatrix}$, $E_i = \begin{bmatrix} E_{ai} & 0 \\ 0 & A_{zi}E_{si} \end{bmatrix}$, $C_i = \begin{bmatrix} 0 & I \end{bmatrix}$

Based on [5.4], the overall fuzzy observer is represented as follows:

$$\dot{\hat{X}}(t) = \sum_{i=1}^{p} \mu_i[A_i\hat{X}(t) + B_i U(t) + E_i \hat{f}(t) + K_i(Y(t) - \hat{Y}(t))],$$

$$\dot{\hat{f}}(t) = \sum_{i=1}^{p} \mu_i L_i(Y - \hat{Y}) = \sum_{i=1}^{p} \mu_i L_i \tilde{Y},$$

$$\hat{Y}(t) = \sum_{i=1}^{p} \mu_i C_i \hat{X}(t) \qquad [5.40]$$

where K_i is the proportional observer gain and L_i is its integral gain for the *i*th observer rule. $Y(t)$ and $\hat{Y}(t)$ are the final output of the fuzzy system and the fuzzy observer, respectively.

5.5.2. Proposed RDFFTC, reference model and stability analysis

In order to establish the conditions for the asymptotic convergence of the observers [5.40], consider the stable linear model without faults described as in Boukhezzar et al. (2006), Kamal et al. (2011, 2012, 2013), Khan et al. (2011) and Chen et al. (1996).

5.5.2.1. Proposed RDFFTC

Based on the analysis given (Chen et al. 1996; Boukhezzar et al. 2006; Kamal et al. 2011, 2012, 2013; Khan et al. 2011), we construct the following fuzzy controller:

$$U(t) = \sum_{i=1}^{p} \mu_i u_i(t) \qquad [5.41]$$

From [5.38], [5.39] and [5.41], we have

$$\dot{X}(t) = \sum_{i=1}^{p} \mu_i (A_i + \Delta A_i) X(t) + BU(t) + Ef(t)$$

$$Y(t) = \sum_{i=1}^{p} \mu_i C_i X(t) \qquad [5.42]$$

Note that B and E are known.

5.5.2.2. Stability analyses of the proposed RDFFTC

The analysis procedures are the same as those in section 5.4.2.3, so the analysis results will be presented without proof. The main result for the global asymptotic stability of a TS fuzzy model with parameter uncertainties, sensor faults and actuator faults are summarized by the following lemma and theorem.

LEMMA 5.1.– *The fuzzy control system as given by [5.42] is stable if B_i is non-singular and the control laws of fuzzy controller of [5.41] are designed as*

$$u_i(t) = B_i^{-1} \Big\{ [He_1(t) + \bar{A}\bar{x}(t) + \bar{B}r(t) - A_i x(t) - S\hat{f}(t) \\ - \frac{e_1(t)\|e_1(t)\|\|P_1\|\|\Delta A_i\|_{max}\|x(t)\|}{e_1(t)^T P_1 e_1(t)} - \frac{e_1(t)\|x(t)\|\|D\|_{max}\|x(t)\|}{e_1(t)^T P_1 e_1(t)} \\ - \frac{e_1(t)\|e_1(t)\|\|P_1\|\|E\|_{max}\|\tilde{f}(t)\|}{e_1(t)^T P_1 e_1(t)}] \Big\} \qquad [5.43]$$

$\|\cdot\|$ denotes the l_2 norm for vectors and l_2 induced norm for matrices, $\|\Delta A_i\| \le \|\Delta A_i\|_{max}$, $\|E\| \le \|E\|_{max}$, $\|D\| \le \|D\|_{max}$ and $H \in \kappa^{n \times n}$ is a stable matrix to be designed and choose S so that $S = E$ and $D = B_{oi}^T B_{oi}$.

THEOREM 5.3.– *If there exist symmetric and positive definite matrices P_{11}, some matrices K_i and L_i, and matrices Z_{qi} and Q_i, such that the following LMEs are satisfied, then the TS fuzzy system [5.42] is asymptotically stabilizable via the TS fuzzy model based output-feedback controller [5.41] and [5.43]*

$$A_i^T P_{11} + P_{11} A_i + (Z_{qi} C_i)^T + (Z_{qi} C_i) = -\sigma I \qquad [5.44]$$

$$(Q_i C_i)^T + (Q_i C_i) = -\sigma I \qquad [5.45]$$

According to the above-mentioned analysis, the procedure for finding the proposed fuzzy FTC controller and the FPIEO observer are summarized as follows:

1) obtain the mathematical model of the hybrid wind-diesel storage system (HWDSS) to be controlled;

2) obtain the fuzzy plant model for the system stated in step (1) by means of a fuzzy modeling method;

3) check if there exists B^{-1} by determining its rank;

4) choose a stable reference model;

5) solve LMEs [5.44] and [5.45] to obtain Q_i, Z_{qi}, P_{11}, K_i and L_i, thus $Z_{qi} = -P_{11} K_i$ and $Q_i = -P_{11} L_i$;

6) construct fuzzy observer [5.40] according to the theorem and fuzzy controller [5.41] according to Lemma 5.1.

By solving [5.44] and [5.45] related to HWDSS, the robustness index σ is shown in Figure 5.13.

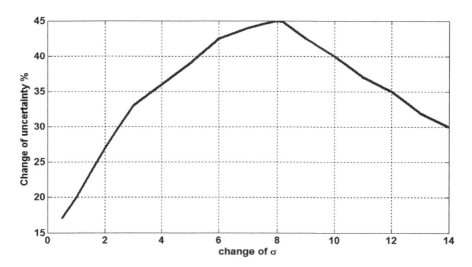

Figure 5.13. *Change of uncertainty with σ*

5.5.3. HWDSS application and simulations and results

In this section, we present the description of the HWDSS and its input and output relationship from the control point of view and TS fuzzy model for HWDSS description.

5.5.3.1. HWDSS description

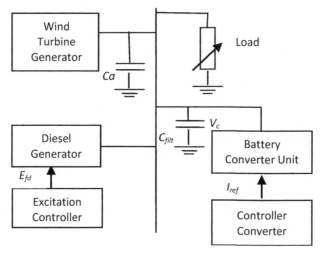

Figure 5.14. *Structural diagram of hybrid wind-diesel storage system*

The HWDSS consists of a horizontal axis, three-bladed, stall regulated WT with a rotor of 16.6 m diameter, and equipped with an IG rated at 55 kW connected to an AC bus-bar in parallel with a diesel-generator set consisting of a 50 kW turbocharged diesel engine (DE) driving a 65 kVA brushless synchronous generator (SG) and an energy storage system. The two generators are connected to a common AC bus-bar. The overall structure of the wind-battery system is shown in Figure 5.14 (Chen et al. 1996; Boukhezzar et al. 2006; Kamal et al. 2011, 2012, 2013; Khan et al. 2011).

The dynamics of the nonlinear **HWDSS** can be characterized by the following equations:

$$\dot{x} = A(x)x(t) + B(x)u(t), \quad y = Cx(t) \qquad [5.46]$$

where $x(t) = [V_b \; \omega_s]^T = [x_1(t) \; x_2(t)]^T$, $u(t) = [E_{fd} \; I_{ref}]^T = [u_1(t) \; u_2(t)]^T$

$$A(x) = \begin{bmatrix} 1 & 1 \\ 0 & 1 \end{bmatrix} \begin{bmatrix} \dfrac{L_f}{\tau L_{md}} q_1(x) & \dfrac{L_f}{\tau L_{md}} q_1(x)(L_{di_{sd}} - r_{ai_{sq}} q_1(x)) \\ \dfrac{P_{ind} - P_{load}}{J_s} q_2(x) & -\dfrac{D_s}{J_s} \end{bmatrix}$$

$$B(x) = \begin{bmatrix} 1 & -V_c q_1(x)/J_s \\ 0 & -V_c q_1(x)/J_s \end{bmatrix}, \quad C = \begin{bmatrix} 1 & 0 \\ 0 & 1 \end{bmatrix}$$

where $q_1(x(t)) = 1/\omega_s$ and $q_2(x(t)) = 1/V_b\omega_s$ are the nonlinear terms, V_c is the AC side voltage of the converter, E_{fd} is the SG field voltage, ω_s is the angular speed, which is proportional to frequency f_r, J_s and D_s are the inertia and frictional damping of SG, respectively, i_{sd} and i_{sq} are the direct and quadrature current components of SG, respectively, L_d and L_f are the stator d-axis and rotor inductances of SG, respectively, L_{md} is the d-axis field mutual inductance, τ is the transient open circuit time constant, r_a is the rotor resistance of SG, P_{ind} is the power of the IG, P_{load} is the power of the load, I_{ref} is the direct current set point, V_b is the bus voltage and C_a is the capacitor bank. Equation [5.46] indicates that the matrices $A(x)$ and $B(x)$ are not fixed, but change as functions of state variables, thus making the model nonlinear.

5.5.3.2. TS fuzzy HWDSS description

We first represent the system [5.46] by a TS fuzzy representation with the angular speed of SG (ω_s) and the bus voltage (V_b) as the measurable premise variables. Consequently, the HWDSS can be represented by a TS fuzzy plant model having four rules. The ith rule can be written as follows ($i = 1,2,3,4$):

Rule i: IF $q_1(t)$ is N_{ji} and $q_2(t)$ is M_{ji}

then $\dot{x}(t) = (A_i + \Delta A_i)x(t) + B_i u(t) + E_{ai} f_a(t)$,

$$y(t) = C_i x(t) + E_{si} f_s(t) \quad j = 1,2; i = 1,2,...,4 \quad [5.47]$$

Referring to [5.36], the fuzzy plant model is given by

$$\dot{x}(t) = \sum_{i=1}^{4} \mu_i [(A_i + \Delta A_i)x(t) + B_i U(t) + E_i f(t)],$$

$$y(t) = \sum_{i=1}^{4} \mu_i [C_i x(t) + E_{is} f_s(t)] \quad [5.48]$$

where $x(t)\in\kappa^{2x1}, U(t)\in\kappa^{2x1}$ are the state vectors and the control input, respectively, ΔA_i represents the system parameters uncertainties but bounded, the elements of ΔA_i randomly achieve the values within 30% of their nominal values corresponding to A_i, which represent the change of parameter uncertainties L_f, L_d, and D_s within 30% of their nominal values and $\Delta B_i = 0$. The minimum and maximum values of $q_1(t)$ and $q_2(t)$ are given by $q_1(t)\in[q_{1min}\ q_{1max}] \in[1/240\ 1/220]$ and $q_2(t)\in[q_{2min}\ q_{2max}]\in[1/3\ 1/0.1]$; the membership functions ($N_i(q_1(t))$, $M_i(q_2(t))$) are related to the uncertain system parameters, which are given by

$$N_1(q_1(t)) = \frac{q_1(t)-q_{1\min}}{q_{1\max}-q_{1\min}}, N_2(q_1(t)) = 1 - N_1(q_1(t)),$$

$$M_1(q_2(t)) = \frac{q_2(t)-q_{2\min}}{q_{2\max}-q_{2\min}}, M_2(q_2(t)) = 1 - M_1(q_2(t)) \quad [5.49]$$

$$A_1 + \Delta A_1 = \begin{bmatrix} 1 & 1 \\ 0 & 1 \end{bmatrix} \begin{bmatrix} \dfrac{L_f + \Delta L_f}{\tau L_{md}} q_{1\max} & \dfrac{L_f + \Delta L_f}{\tau L_{md}} q_{1\max}((L_d + \Delta L_d)i_{sd} - r_a i_{sq} q_{1\max}) \\ \dfrac{P_{ind} - P_{load}}{J_s} q_{2\min} & -\dfrac{D_s + \Delta D_s}{J_s} \end{bmatrix}$$

$$A_2 + \Delta A_2 = \begin{bmatrix} 1 & 1 \\ 0 & 1 \end{bmatrix} \begin{bmatrix} \dfrac{L_f + \Delta L_f}{\tau L_{md}} q_{1\max} & \dfrac{L_f + \Delta L_f}{\tau L_{md}} q_{1\max}((L_d + \Delta L_d)i_{sd} - r_a i_{sq} q_{1\max}) \\ \dfrac{P_{ind} - P_{load}}{J_s} q_{2\max} & -\dfrac{D_s + \Delta D_s}{J_s} \end{bmatrix}$$

$$A_3 + \Delta A_3 = \begin{bmatrix} 1 & 1 \\ 0 & 1 \end{bmatrix} \begin{bmatrix} \dfrac{L_f + \Delta L_f}{\tau L_{md}} q_{1\min} & \dfrac{L_f + \Delta L_f}{\tau L_{md}} q_{1\min}((L_d + \Delta L_d)i_{sd} - r_a i_{sq} q_{1\min}) \\ \dfrac{P_{ind} - P_{load}}{J_s} q_{2\min} & -\dfrac{D_s + \Delta D_s}{J_s} \end{bmatrix}$$

$$A_4 + \Delta A_4 = \begin{bmatrix} 1 & 1 \\ 0 & 1 \end{bmatrix} \begin{bmatrix} \dfrac{L_f + \Delta L_f}{\tau L_{md}} q_{1\min} & \dfrac{L_f + \Delta L_f}{\tau L_{md}} q_{1\min}((L_d + \Delta L_d)i_{sd} - r_a i_{sq} q_{1\min}) \\ \dfrac{P_{ind} - P_{load}}{J_s} q_{2\max} & -\dfrac{D_s + \Delta D_s}{J_s} \end{bmatrix}$$

$$B_1 = B_2 = \begin{bmatrix} 1 & -\dfrac{V_c}{J_s} q_{1\max} \\ 0 & -\dfrac{V_c}{J_s} q_{1\max} \end{bmatrix}, \quad B_3 = B_4 = \begin{bmatrix} 1 & -\dfrac{V_c}{J_s} q_{1\min} \\ 0 & -\dfrac{V_c}{J_s} q_{1\min} \end{bmatrix},$$

$$E_{ai} = \begin{bmatrix} 1 & 1 \\ 0.1 & 0.1 \end{bmatrix} \quad E_{si} = \begin{bmatrix} 10 & 1 \\ 0.1 & 0.01 \end{bmatrix}$$

To define the state Z, we choose $A_{iz} = 20 \times I$, where I is the identity matrix. Consider $H = \begin{bmatrix} -4 & -4 \\ 0 & -1 \end{bmatrix}$ is a stable matrix. The stable reference model is chosen as follows:

$$\bar{A} = \begin{bmatrix} -4 & -4 \\ 0 & -4 \end{bmatrix}, \bar{B} = \begin{bmatrix} 1 & 1 \\ 0 & 1 \end{bmatrix}, \bar{C} = \begin{bmatrix} 1 & 0 \\ 0 & 1 \end{bmatrix}$$

Solve LMEs [5.44] and [5.45] give the computation of the matrices K_i and L_i, construct the fuzzy FTC [5.41] and construct the fuzzy observer based on the theorem [5.40].

5.5.3.3. Simulation studies

The proposed RDFFTC for the HWDSS [5.46] is tested and we study the responses for the HWDSS subject to sensor faults, actuator faults and parameter uncertainties. The proposed controller is tested for same random profiles of wind speed to prove the effectiveness of the proposed algorithm. The actuator fault $f_a(t) = [f_{a1}\ f_{a2}]^T$ are modeled as follows:

$$f_{a1}(t) = \begin{cases} 0 & t < 13.34\,\text{sec} \\ 3\sin(\pi t) & t \geq 13.34\,\text{sec} \end{cases} \quad f_{a2}(t) = \begin{cases} 0 & t \leq 13.34\,\text{sec} \\ \sin(\pi t) & t \geq 13.34\,\text{sec} \end{cases} \quad [5.50]$$

where $f_{a1}(t)$ is the excitation field voltage actuator fault of the SG and the direct-current set point actuator fault of the converter is $f_{a2}(t)$. The sensor faults $f_s(t) = [f_{s1}\ f_{s2}]^T$ are modeled as follows:

$$f_{s1}(t) = \begin{cases} 0 & t < 22.22\,s \\ 10\sin(\pi t) & t \geq 22.22\,s \end{cases}, \quad f_{s2}(t) = \begin{cases} 0 & t < 22.22\,s \\ 5\sin(\pi t) & t \geq 22.22\,s \end{cases} \quad [5.51]$$

where f_{s1} is the bus voltage sensor and f_{s2} is the generator speed sensors. Figure 5.15 shows the actuator fault (top) and sensor faults (bottom) and their estimations based on [5.50] and [5.51], respectively. The frequency and bus voltage of the system (dashed line), observer states (dotted line) and reference states (solid line) are shown in Figures 5.16 and 5.17, respectively. When the wind speed is varying, the produced power curve takes almost the wind speed curve as shown in Figure 5.18, but there is only spike when the fault is detected at 13.34 and 22.22 seconds; it is clear that a 35% increase is obtained at the maximum value compared with Chedid (2000).

It can be seen from the simulation results that there is a good tracking between the states of the nonlinear HWDSS subject to the norm-bounded parametric uncertainties, sensor faults and actuator faults and the reference model. Thus, the TS fuzzy model based controller through fuzzy observer is robust against norm-bounded parametric uncertainties, sensor faults and actuator faults. Comparing the results of the proposed algorithm with that given in the previous algorithms, it can be seen that the proposed controller has the following advantages: (i) it is stable over a wide range of uncertainty up to 30% compared with Chedid (2000), and when the actuators and sensors become faulty at any time; (ii) the generated power is increased and bus voltage is nearly constant compared with Chedid (2000).

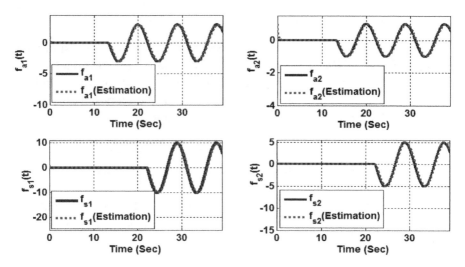

Figure 5.15. *Actuator faults ($f_{a1}(t)$ and $f_{a2}(t)$) and their estimations (top) and sensor faults ($f_{s1}(t)$ and $f_{s2}(t)$) and their estimations (bottom)*

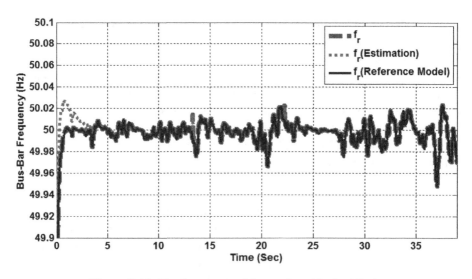

Figure 5.16. *The frequency of the system (dashed line), observer (dotted line) and reference model (solid line)*

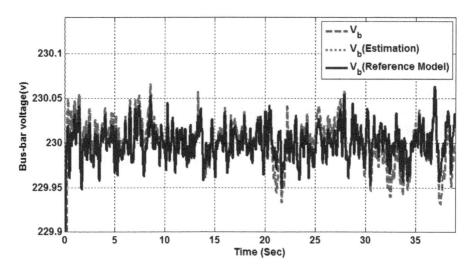

Figure 5.17. *The bus voltage of the system (dashed line), observer (dotted line) and reference model (solid line)*

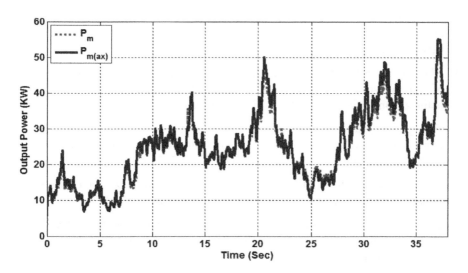

Figure 5.18. *Produced power with the proposed control law*

5.6. Conclusion

An augmented TS fuzzy plant model has been proposed to model the nonlinear plant subject to large parameter uncertainties, sensor faults and actuator faults. Based on this augmented TS fuzzy plant model, three different methods to design the fuzzy fault-tolerant control has been proposed to tackle this nonlinear system. A design procedure of fuzzy FTCs has been developed. The stability and robustness of the fuzzy fault-tolerant control systems have been investigated based on the results of Boukhezzar et al. (2006), Kamal et al. (2011, 2012, 2013), Khan et al. (2011) and Chen et al. (1996).

An application example on stabilizing a WES with sensor faults, actuator faults and parameter uncertainties has been given to illustrate the design procedure and merits of the proposed fuzzy FTC.

5.7. References

Abo-Khalil, A.G. and Dong-Choon, L. (2008). MPPT control of wind generation systems based on estimated wind speed using SVR. *IEEE Trans. Ind. Electron.*, 55(3), 1489–1490.

Aggarwal, V., Patidar, R.K., Patki, P. (2010). A novel scheme for rapid tracking of maximum power point in wind energy generation systems. *IEEE Trans. Energy Convers.*, 25(1), 228–236.

Boukhezzar, B., Siguerdidjane, H., Maureen Hand, M. (2006). Nonlinear control of variable-speed wind turbines for generator torque limiting and power optimization. *J. Sol. Energy Eng.*, 128(4), 516–530.

Camblong, H., Martinez de Alegria, I., Rodriguez, M., Abad, G. (2006). Experimental evaluation of wind turbines maximum power point tracking controllers. *Energy Convers. Manage.*, 47(18–19), 2846–2858.

Chedid, R.B., Karaki, S.H., El-Chamali, C. (2000). Adaptive fuzzy control for wind diesel weak power systems. *IEEE Trans. Energy Convers.*, 15(1), 71–78.

Chen, W. and Saif, M. (2007). Design of a TS based fuzzy nonlinear unknown input observer with fault diagnosis applications. In *American Control Conference*. July 9–13, New York.

Chen, J., Patton, R., Zhang, H. (1996). Design of unknown input observers and robust fault detection filters. *Int. J. Control*, 63, 85–105.

Connor, B., Iyer, S.N., Leithead, W.E., Grimble, M.J. (1992). Control of a horizontal axis wind turbine using H∞ control. In *Proceedings of the First IEEE Conference on Control Applications*. September 13–16, Dayton.

Datta, R. and Ranganathan, V.T. (2003). A method of tracking the peak power points for a variable speed wind energy conversion system. *IEEE Trans. Energy Convers.*, 18(1), 163–168.

Galdi, V., Piccolo, A., Siano, P. (2009). Exploiting maximum energy from variable speed wind power generation system by using an adaptive Takagi–Sugeno–Kang fuzzy model. *Energy Convers. Manage.*, 50, 413–420.

Guo, S.X. (2010). Robust reliability as a measure of stability of controlled dynamic systems with bounded uncertain parameters. *J. Vib. Control*, 16(9), 1351–1368.

Jayaram, S. and Johnson, R.W. (2010). Robust fault-tolerant control architecture-actuator fault detection and reconfiguration. *Control and Intelligent Systems*, 38, 120–128.

Kamal, E. and Aitouche, A. (2013). Design of maximum power fuzzy controller for PV systems based on the LMI-based stability. Intelligent systems in technical and medical diagnostics. *Advances in Intelligent Systems and Computing*, 230, 77–88.

Kamal, E., Aitouche, A., Bayart, M. (2011). Robust control of wind energy conversion systems. In *2011 International Conference on Communications, Computing and Control Applications (CCCA)*. March 3–5, Hammamet.

Kamal, E., Aitouche, A., Ghorbani, R., Bayart, M. (2012a). Intelligent control of WECS subject to parameter uncertainties, actuator and sensor faults. *Control and Intelligent Systems*, 40(3), 160.

Kamal, E., Aitouche, A., Ghorbani, R., Bayart, M. (2012b). Robust fuzzy logic control of wind energy conversion systems with unknown inputs. *International Journal of Power and Energy Systems*, 32(2), 71.

Kamal, E., Aitouche, A., Ghorbani, R., Bayart, M. (2012c). Fault tolerant control of wind energy system subject to actuator faults and time varying parameters. In *2012 20th Mediterranean Conference on Control & Automation (MED)*, 1177–1182. July 3–6, Barcelona.

Kamal, E., Aitouche, A., Ghorbani, R., Bayart, M. (2012d). Unknown input observer with fuzzy fault tolerant control for wind energy system. *IFAC Proceedings Volumes*, 45(20), 946–951.

Kamal, E., Aitouche, A., Kuzmych, O. (2013). Robust fuzzy controller for photovoltaic maximum power point tracking. In *21st Mediterranean Conference on Control and Automation*. June 25–28, Chania.

Khan, S.A. and Hossain, M.I. (2011). Intelligent control based maximum power extraction strategy for wind energy conversion system. In *24th Canadian Conference on Electrical and Computer Engineering (CCECE)*. May 11, Niagara Falls, Ontario.

Koutroulis, E. and Kalaitzakis, K. (2006). Design of a maximum power tracking system for wind-energy-conversion applications. *IEEE Trans. Ind. Electron.*, 53(2), 486–494.

Leith, D.J. and Leithead, W.E. (1999). Survey of gain scheduling analysis design. *Int. J. Control*, 73(11), 1001–1025.

Marx, B., Koenig, D., Georges, D. (2004). A robust fault tolerant control for descriptor systems. *IEEE Trans. Autom. Control*, 49(10), 1869–1875.

Masoud, B.S., Kazerani, M., Aplevich, J.D. (2009). Maximum power tracking control for a wind turbine system including a matrix converter. *IEEE Trans. Energy Convers.*, 24(3), 705–713.

Mohamed, A.Z., Eskander, M.N., Ghali, F.A. (2001). Fuzzy logic control based maximum power tracking of a wind energy system. *Renewable Energy*, 23(2), 235–245.

Mufeed, M.M., Jiang, J., Zhang, Z. (2003). *Active Fault Tolerant Control Systems: Stochastic Analysis and Synthesis*. Springer-Verlag, Berlin/Heidelberg.

Niemann, H. and Stoustrup, J. (2005). Passive fault tolerant control of a double inverted pendulum – A case study. *Control Eng. Pract.*, 13(8), 1047–1059.

Ocampo-Martinez, C., De Dona, J., Seron, M. (2010). Actuator fault-tolerant control based on set separation. *Int. J. Adapt. Control Signal Process.*, 24(12), 1070–1090.

Patton, R. (1997). Fault-tolerant control systems. In *The 1997 Situation, 3rd IFAC Symposium on Fault Detection, Supervision and Safety of Technical Processes*. IFAC, August, Kingston.

Rocha, R., Filho, L.S.M., Bortolus, M.V. (2005). Optimal multivariable control for wind energy conversion system – A comparison between H2 and H∞ controllers. In *Proceedings of the 44th IEEE Conference on Decision & Control*. December 12–15, Seville.

Seron, M., Zhuo, X., De Donà, J., Martinez, J. (2008). Multisensor switching control strategy with fault tolerance guarantees. *Automatica*, 44(1), 88–97.

Shtessel, Y., Buffington, J., Banda, S. (2002). Tailless aircraft flight control using multiple time scale reconfigurable sliding modes. *IEEE Trans. Control Systems Tech.*, 10(2), 288–296.

Stilwell, D.J. and Rugh, W.J. (1997). Interpolation of observer state feedback controllers for gain scheduling. *IEEE Trans. Autom. Control*, 44(6), 1225–1229.

Tanaka, K. and Sugeno, M. (1992). Stability analysis and design of fuzzy control systems. *Fuzzy Sets Syst.*, 45(2), 135–156.

Tong, S. and Han-Hiong, L. (2002). Observer-based robust fuzzy control of nonlinear systems with parametric uncertainties. *Fuzzy Sets Syst.*, 131, 165–184.

Whei-Min, L. and Chih-Ming, H. (2010). Intelligent approach to maximum power point tracking control strategy for variable-speed wind turbine generation system. *Energy*, 35(6), 2440–2447.

Yao, X., Liu S., Shan, G., Xing, Z., Guo, C., Yi, C. (2009). LQG controller for a variable speed pitch regulated wind turbine. In *International Conference on Intelligent Human-Machine Systems and Cybernetics*. IEEE, August 26–27, Hangzhou.

Zhang, Y. and Jiang, J. (2008). Bibliographical review on reconfigurable fault tolerant control systems. *Annual Reviews in Control*, 32(2), 229–252.

Zuo, Z., Ho, D.W.C., Wang, Y. (2010). Fault tolerant control for singular systems with actuator saturation and nonlinear perturbation. *Automatica*, 46(3), 569–576.

List of Authors

Dhaker ABBES
L2EP – ULR 2697
Junia – Grande École d'Ingénieurs
HEI – École des hautes études
d'ingénieur
Lille
France

Abdel AITOUCHE
CRIStAL UMR CNRS 9189
Junia – Grande École d'Ingénieurs
Lille
France

Khmais BACHA
LISIER
ENSIT
Université de Tunis
Tunisia

Monia BEN KHADER BOUZID
LSE
ENIT
Université Tunis El Manar
and
ENICarthage
Université de Carthage
Tunisia

Gérard CHAMPENOIS
LIAS
Université de Poitiers
France

Sérgio CRUZ
Department of Electrical and
Computer Engineering
and
Instituto de Telecomunicações
University of Coimbra
Portugal

Pedro GONÇALVES
MARC
McMaster University
Hamilton
Canada

Elkhatib KAMAL
LS2N
École Centrale Nantes
France
and
Department of Industrial
Electronics and Control
Engineering
University Manoufia
Egypt

Youssef KRAIEM
L2EP – ULR 2697
Junia – Grande École d'Ingénieurs
HEI – École des hautes études
d'ingénieur
Lille
France

Walid TOUTI
LISIER
ENSIT
Université de Tunis
Tunisia

Index

A, D, E

adaptive droop control, 139
average current, 15, 17, 18, 24, 28, 33, 36
diode rectifier, 2–4, 8, 11, 29, 37
discrete
   cosine transform (DCT), 90, 111, 112, 114
   sine transform (DST), 90, 111, 113, 114
extended park vector approach (EPVA), 106, 108–111

F, G

fault(s)
   actuator, 161–167, 169, 171, 173–176, 178–181, 183, 184, 186, 187, 189, 195, 196, 198
   current sensor, 76, 79, 80
   detection and diagnosis (FDD), 160, 162
   detection, discrimination, location
      detection, 2, 4, 32, 35
      discrimination, 32, 33, 35, 37
      location, 4, 17, 18, 32, 35–37
   diagnosis, 71, 76, 77, 81

fault-tolerant control (FTC), 160–168, 170, 171, 174–186, 188–190, 195, 198
gearbox fault detection, 90, 91, 116
inter-turn short-circuit (ITSCF), 2, 4–6, 8, 11–14, 16–18, 24–28, 31–37
interturn short-circuit (ITSC), 73–75
open-circuit diode (OCDF), 2, 4, 7, 8, 11, 12, 14, 15, 19–25, 29–37
open-phase (OPFs), 40, 76
permanent magnet (PM), 40, 43, 71, 73, 74, 78, 79, 81
sensor, 160–164, 166, 167, 171, 175, 176, 178–181, 183, 184, 186, 187, 189, 195–198
signature, 110
frequency regulation, 122
fuzzy
   logic, 122, 123, 130, 132–136, 139, 141, 143, 144, 147–150, 154
   observer, 164, 165, 174, 178, 182, 187, 188, 190, 195
   system(s), 160, 162, 163, 167, 168, 170, 178, 179, 181, 186, 188, 190

gear failure, 90
grid-connected mode, 144, 150

H, I, L

high-resistance connections (HRCs), 40, 76
hybrid storage system (HSS), 121, 122, 124, 128, 130, 132, 133, 135, 136, 153
induction machine, 89–91, 93, 94, 97, 102, 106, 108, 116
islanding detection, 139, 141, 142
linear matrix inequalities (LMIs), 162

M, N, O

meshing frequency, 91, 96, 98, 102, 104
model predictive control (MPC), 40, 51, 54–56, 81
motor current signature analysis (MCSA), 89, 90, 101, 103, 106, 109, 110, 115, 116
multiphase drives, 71, 80, 81
negative
  sequence current (NSC), 4, 5, 11, 12, 15, 24, 26, 28, 30, 31, 36, 37
  sequence voltage (NSV), 2, 4, 5, 7, 10, 12, 14–17, 19, 21, 23–34, 36, 37

P, R, S

parallel distributed computing (PDC), 159, 165, 167, 178
permanent
  magnet synchronous generator (PMSG), 2–5, 8, 9, 12, 14, 16, 17, 19, 20, 25–29, 33, 37, 39–44, 47, 48, 50–52, 54–56, 66, 68, 69, 71, 74, 75, 77, 80, 81
power
  grid, 120, 122–124, 128, 130, 132, 138, 139, 141, 144, 146, 147, 150, 151, 154
  management, 121, 128, 130, 132, 139, 147, 150, 154
renewable distributed generator, 123, 153

T, V, W

Takagi-Sugeno model (TS), 159, 161–165, 167, 168, 171, 178–182, 186, 187, 189–191, 193, 195, 197, 198
voltage regulation/control, 122, 145, 146
wind
  energy system (WES), 161, 163, 166, 171, 174, 175–179, 183, 186, 198
  turbines (WTs), 39

Other titles from

in

Energy

2023

ALVAREZ-HERAULT Marie-Cécile, GOUIN Victor, CHARDIN-SEGUI Trinidad, MALOT Alain, COIGNARD Jonathan, RAISON Bertrand, COULET Jérôme
Distribution System Planning: Evolution of Methodologies and Digital Tools for Energy Transition

2022

ABOUELATTA Mohamed, SHAKER Ahmed, GONTRAND Christian
Smart Power Integration

BOUTAUD Benoit
Energy Autonomy: From the Notion to the Concepts

2021

ROBYNS Benoît, DAVIGNY Arnaud, FRANÇOIS Bruno, HENNETON Antoine, SPROOTEN Jonathan
Electricity Production from Renewable Energies (2nd edition)

2020

BOISGIBAULT Louis, AL KABBANI Fahad
Energy Transition in Metropolises, Rural Areas and Deserts

SOUALHI Abdenour, RAZIK Hubert
Electrical Systems 1: From Diagnosis to Prognosis
Electrical Systems 2: From Diagnosis to Prognosis

2019

BENALLOU Abdelhanine
Energy Transfers by Convection
(Energy Engineering Set – Volume 3)
Energy Transfers by Radiation
(Energy Engineering Set – Volume 4)
Mass Transfers and Physical Data Estimation
(Energy Engineering Set – Volume 5)

LACHAL Bernard
Energy Transition

ROBYNS Benoît, DAVIGNY Arnaud, BARRY Hervé, KAZMIERCZAK Sabine, SAUDEMONT Christophe, ABBES Dhaker, FRANÇOIS Bruno
Electrical Energy Storage for Buildings in Smart Grids

2018

BENALLOU Abdelhanine
Energy and Mass Transfers: Balance Sheet Approach and Basic Concepts
(Energy Engineering Set – Volume 1)
Energy Transfers by Conduction
(Energy Engineering Set – Volume 2)

JEMEÏ Samir
Hybridization, Diagnostic and Prognostic of Proton Exchange Membrane Fuel Cells: Durability and Reliability

RUFFINE Livio, BROSETA Daniel, DESMEDT Arnaud
Gas Hydrates 2: Geoscience Issues and Potential Industrial Applications

VIALLET Virginie, FLEUTOT Benoit
Inorganic Massive Batteries
(Energy Storage – Batteries, Supercapacitors Set – Volume 4)

2017

BROSETA Daniel, RUFFINE Livio, DESMEDT Arnaud
Gas Hydrates 1: Fundamentals, Characterization and Modeling

LA SCALA Massimo
From Smart Grids to Smart Cities: New Challenges in Optimizing Energy Grids

MOLINA Géraldine, MUSY Marjorie, LEFRANC Margot
Building Professionals Facing the Energy Efficiency Challenge

SIMON Patrice, BROUSSE Thierry, FAVIER Frédéric
Supercapacitors Based on Carbon or Pseudocapacitive Materials
(Energy Storage – Batteries, Supercapacitors Set – Volume 3)

2016

ALLARD Bruno
Power Systems-on-Chip: Practical Aspects of Design

ANDRÉ Michel, SAMARAS Zissis
Energy and Environment

DUFOUR Anthony
Thermochemical Conversion of Biomass for the Production of Energy and Chemicals

2015

CROGUENNEC Laurence, MONCONDUIT Laure, DEDRYVÈRE Rémi
Electrodes for Li-ion Batteries
(Energy Storage – Batteries, Supercapacitors Set – Volume 2)

LEPRINCE-WANG Yamin
Piezoelectric ZnO Nanostructure for Energy Harvesting
(Nanotechnologies for Energy Recovery Set – Volume 1)

ROBYNS Benoît, FRANÇOIS Bruno, DELILLE Gauthier,
SAUDEMONT Christophe
Energy Storage in Electric Power Grids

ROSSI Carole
*Al-based Energetic Nanomaterials
(Nanotechnologies for Energy Recovery Set – Volume 2)*

TARASCON Jean-Marie, SIMON Patrice
*Electrochemical Energy Storage
(Energy Storage – Batteries, Supercapacitors Set – Volume 1)*

2013

LALOUI Lyesse, DI DONNA Alice
Energy Geostructures: Innovation in Underground Engineering

2012

BECKERS Benoit
Solar Energy at Urban Scale

ROBYNS Benoît, DAVIGNY Arnaud, FRANÇOIS Bruno, HENNETON Antoine,
SPROOTEN Jonathan
Electricity Production from Renewable Energies

2011

GAO Fei, BLUNIER Benjamin, MIRAOUI Abdellatif
Proton Exchange Membrane Fuel Cell Modeling

MULTON Bernard
Marine Renewable Energy Handbook

2010

BRUNET Yves
Energy Storage

2009

SABONNADIÈRE Jean-Claude
Low Emission Power Generation Technologies and Energy Management

SABONNADIÈRE Jean-Claude
Renewable Energy Technologies

Printed and bound by CPI Group (UK) Ltd, Croydon, CR0 4YY
18/09/2023

08117277-0002